移动互联网开发技术丛书

React
全栈式实战开发入门
微课视频版

吴 胜 编著

清华大学出版社
北京

内 容 简 介

React（React.js）作为三大前端开发框架之一，有着广泛的应用。本书由浅入深、循序渐进地介绍React的应用开发。本书共分三部分，共15章。第一部分基础篇，包括第1～7章，内容涉及React简介与开发基础、React组件、React事件处理、React条件渲染、列表和key、React状态管理、React表单、React组件的组合和继承。第二部分高阶篇，包括第8～10章，内容涉及React使用Hook增强组件、React基础原理和高级指引、React应用开发的工具。第三部分实战篇，包括第11～15章，内容涉及React与Redux的整合开发、React与Spring Boot的整合开发、React与Python框架的整合开发、React与Go的整合开发，案例。

本书适合作为全国高等院校前端开发类课程的教材，也可供从事React应用开发和Web前端开发人员参考。

本书封面贴有清华大学出版社防伪标签。无标签者不得销售。
版权所有，侵权必究。举报：010-62782989，beiqinquan@tup.tsinghua.edu.cn。

图书在版编目（CIP）数据

React全栈式实战开发入门：微课视频版 / 吴胜编著. —北京：清华大学出版社，2023.1（2023.11重印）
（移动互联网开发技术丛书）
ISBN 978-7-302-61559-0

Ⅰ.①R… Ⅱ.①吴… Ⅲ.①移动终端–应用程序–程序设计 Ⅳ.①TN929.53

中国版本图书馆CIP数据核字（2022）第143777号

责任编辑：陈景辉
封面设计：刘　键
责任校对：韩天竹
责任印制：宋　林

出版发行：清华大学出版社
网　　址：http://www.tup.com.cn, http://www.wqbook.com
地　　址：北京清华大学学研大厦A座　　邮　编：100084
社 总 机：010-83470000　　邮　购：010-62786544
投稿与读者服务：010-62776969, c-service@tup.tsinghua.edu.cn
质 量 反 馈：010-62772015, zhiliang@tup.tsinghua.edu.cn
课 件 下 载：http://www.tup.com.cn, 010-83470236

印 装 者：三河市铭诚印务有限公司
经　　销：全国新华书店
开　　本：185mm×260mm　　印　张：17　　字　数：417千字
版　　次：2023年1月第1版　　印　次：2023年11月第2次印刷
印　　数：1501～2300
定　　价：79.90元

产品编号：096389-01

前言
PREFACE

React（或称为 React.js）作为三大前端开发框架之一，有着广泛的应用。本书由浅入深、循序渐进地介绍 React 的应用开发，不包含 React Native 的应用开发。在介绍 React 应用开发的基础上，考虑到开发时前后端分离的需求和 React 的灵活性，本书还介绍了不同的框架（语言）与 React 的整合开发，如 Spring Boot、Python 框架（Django 和 Flask）、Go 等与 React 的整合开发。结合不同框架（语言）与 React 的整合开发还介绍了 fetch、axios、把前端打包到后端、XMLHttpRequest 等方式的前后端整合开发。

本书主要内容

本书分为三部分，共 15 章。

第一部分基础篇，包括第 1~7 章。

第 1 章 React 简介与开发基础，内容包括 React 简介、React 应用开发的简单示例、JSX 应用开发入门。

第 2 章 React 组件，内容包括 React 组件概述、函数组件和类组件的应用开发、组件参数和组合组件、组件的分解和组合、组件的生命周期。

第 3 章 React 事件处理，内容包括 React 事件处理概述、鼠标事件处理、焦点事件处理、键盘事件处理和图像事件处理。

第 4 章 React 条件渲染、列表和 key，内容包括 React 条件渲染、列表和 key 概述、条件渲染的应用开发、列表的应用开发、key 的应用开发、列表和 key 的综合应用。

第 5 章 React 状态管理，内容包括 React 状态管理概述、状态的基础应用、状态的提升应用。

第 6 章 React 表单，内容包括 React 表单概述、表单组件和 ref。

第 7 章 React 组件的组合和继承，内容包括 React 组件的组合和继承概述、带样式的组合组件、页面布局、特例关系组合和类组合。

第二部分高阶篇，包括第 8~10 章。

第 8 章 React 使用 Hook 增强组件，内容包括 Hook 概述、State Hook 的应用、State Hook 的综合应用、Effect Hook 的应用和其他 Hook 的应用。

第 9 章 React 基础原理和高级指引，内容包括 React 基础原理、React 应用开发的一般步骤、React 片段、context、高阶组件、ref 转发、portal、ref 和 DOM、Web Component、render props、错误边界和测试。

第 10 章 React 应用开发的工具，内容包括包管理器、安装 React、编译器和编辑器、构建工具、服务器端渲染工具和 React Router。

第三部分实战篇，包括第 11~15 章。

第 11 章 React 与 Redux 的整合开发，内容包括 React 和 Redux 概述、计数器的开发、待办事项管理小工具的开发。

第 12 章 React 与 Spring Boot 的整合开发，内容包括 Spring Boot 简介、Spring Boot 作为后端的开发、React 作为前端的开发。

第 13 章 React 与 Python 框架的整合开发，内容包括 React 与 Django 的整合开发、React 与 Flask 的整合开发。

第 14 章 React 与 Go 的整合开发，内容包括 Go 作为后端的开发、React 作为前端的开发。

第 15 章案例——实现一个简易的员工信息管理系统，内容包括 Spring Boot 作为后端的开发、React 作为前端的开发。

本书特色

（1）内容新。本书使用的 React 版本是 18.2.0 版，涵盖新内容（如自动批处理）。

（2）易理解。本书避免对官方文档的简单引用，按照学习的先后顺序和开发步骤由浅入深地编排知识点，适合读者自学，同时也能满足全国高等院校教学的需要。

（3）全栈式。本书包括官方文档的大部分内容，在介绍 React 应用开发的基础上，考虑到开发时前后端分离的需求和 React 的灵活性，本书实战演练了采用 Spring Boot、Python 框架、Go 等不同技术栈和 React 整合开发的案例。

（4）示例多。实战案例丰富，涵盖 48 个知识点示例、6 个整合开发案例、1 个完整项目案例。

配套资源

为便于教与学，本书配有微课视频（210 分钟）、源代码、教学课件、教学大纲、教学进度表、习题答案。

（1）获取微课视频方式：先刮开并扫描本书封底的文泉云盘防盗码，再扫描书中相应的视频二维码，观看教学视频。

（2）获取源代码方式：先刮开并扫描本书封底的文泉云盘防盗码，再扫描下方二维码，即可获取。

源代码　　源代码使用说明

（3）其他配套资源可以扫描本书封底的"书圈"二维码，关注后回复本书书号，即可下载。

读者对象

本书适合作为全国高等院校前端开发类课程的教材,也可供从事 React 应用开发和 Web 前端开发人员参考。

本书的主要内容参考了 React 官方文档,在参考文献已经列出,在此向 React 开发者和官方文档的作者表示衷心的感谢和深深的敬意。本书的编写还参考了诸多同行的相关资料,在此表示衷心的感谢。

限于个人水平和时间仓促,书中难免存在疏漏之处,欢迎读者批评指正。

编者
2022 年 10 月

目录

第一部分 基础篇

第 1 章 React 简介与开发基础 ... 3

- 1.1 React 简介 ... 3
  - 1.1.1 React 的定义 ... 3
  - 1.1.2 React 的特点 ... 3
  - 1.1.3 React 的发展简史 ... 4
- 1.2 React 应用开发的简单示例 ... 5
  - 1.2.1 单个 HTML 文件应用 React 的示例 ... 5
  - 1.2.2 元素渲染说明 ... 8
  - 1.2.3 两个文件应用 React 的示例 ... 9
  - 1.2.4 应用 React 的示例对比分析 ... 10
- 1.3 JSX 应用开发入门 ... 11
  - 1.3.1 JSX 说明 ... 11
  - 1.3.2 JSX 综合应用示例 ... 11
  - 1.3.3 JSX 综合运行效果 ... 14
- 习题 1 ... 14

第 2 章 React 组件 ... 15

- 2.1 React 组件概述 ... 15
  - 2.1.1 组件和自定义组件 ... 15
  - 2.1.2 函数组件和类组件 ... 16
- 2.2 函数组件和类组件的应用开发 ... 16
  - 2.2.1 开发示例 ... 16
  - 2.2.2 运行效果 ... 17
- 2.3 组件参数和组合组件 ... 18
  - 2.3.1 说明 ... 18

  2.3.2 开发示例 18
  2.3.3 运行效果 21
2.4 组件的分解和组合 21
  2.4.1 说明 21
  2.4.2 开发示例 22
  2.4.3 运行效果 24
2.5 组件的生命周期 24
  2.5.1 概述 24
  2.5.2 constructor()方法 25
  2.5.3 componentDidMount()方法 25
  2.5.4 componentDidUpdate()方法 25
  2.5.5 componentWillUnmount()方法 25
  2.5.6 开发示例 26
  2.5.7 运行效果 27
习题 2 28

第 3 章 React 事件处理 29

3.1 React 事件处理概述 29
  3.1.1 事件 29
  3.1.2 合成事件 30
  3.1.3 支持的事件类型 30
3.2 鼠标事件处理 31
  3.2.1 开发示例 31
  3.2.2 运行效果 34
3.3 焦点事件处理 36
  3.3.1 开发示例 36
  3.3.2 运行效果 38
3.4 键盘事件处理 39
  3.4.1 开发示例 39
  3.4.2 运行效果 40
3.5 图像事件处理 41
  3.5.1 开发示例 41
  3.5.2 运行效果 43
习题 3 44

第 4 章 React 条件渲染、列表和 key 45

4.1 React 条件渲染、列表和 key 概述 45
  4.1.1 条件渲染 45
  4.1.2 列表 45

4.1.3　key 46
　4.2　条件渲染的应用开发 46
　　　4.2.1　开发示例 46
　　　4.2.2　运行效果 51
　4.3　列表的应用开发 52
　　　4.3.1　开发示例 52
　　　4.3.2　运行效果 54
　4.4　key 的应用开发 55
　　　4.4.1　开发示例 55
　　　4.4.2　运行效果 58
　4.5　列表和 key 的综合应用 58
　　　4.5.1　开发示例 58
　　　4.5.2　运行效果 60
　习题 4 61

第 5 章　React 状态管理 62

　5.1　React 状态管理概述 62
　　　5.1.1　state 62
　　　5.1.2　setState()方法 63
　　　5.1.3　forceUpdate()方法 64
　　　5.1.4　状态提升 64
　5.2　状态的基础应用 64
　　　5.2.1　开发示例 64
　　　5.2.2　运行效果 70
　5.3　状态的提升应用 71
　　　5.3.1　开发示例 71
　　　5.3.2　运行效果 75
　习题 5 77

第 6 章　React 表单 78

　6.1　React 表单概述 78
　　　6.1.1　表单 78
　　　6.1.2　受控组件 78
　　　6.1.3　非受控组件 79
　6.2　表单组件 79
　　　6.2.1　开发示例 79
　　　6.2.2　运行效果 86
　6.3　ref 87
　　　6.3.1　开发示例 87

    6.3.2 运行效果 89

 习题 6 91

第 7 章 React 组件的组合和继承 92

 7.1 React 组件的组合和继承概述 92
   7.1.1 组合 92
   7.1.2 继承 92
 7.2 带样式的组合组件 93
   7.2.1 引入包、样式和功能文件 93
   7.2.2 定义样式 93
   7.2.3 定义功能 94
   7.2.4 带样式组件综合应用的运行效果 95
 7.3 页面布局 96
   7.3.1 定义样式和功能 96
   7.3.2 运行效果 98
 7.4 特例关系组合 99
   7.4.1 定义样式和功能 99
   7.4.2 运行效果 100
 7.5 类组合 101
   7.5.1 定义样式和功能 101
   7.5.2 运行效果 103
 习题 7 104

第二部分 高阶篇

第 8 章 React 使用 Hook 增强组件 107

 8.1 Hook 概述 107
   8.1.1 Hook 107
   8.1.2 Hook API 108
   8.1.3 自定义 Hook 109
   8.1.4 Hook 的使用规则 109
 8.2 State Hook 的应用 110
   8.2.1 创建项目 reactjsbook 110
   8.2.2 修改文件 index.js 113
   8.2.3 创建组件 113
   8.2.4 运行项目 reactjsbook 114
   8.2.5 useState()函数的应用说明 115
   8.2.6 State Hook 的等价实现 115
 8.3 State Hook 的综合应用 116

		8.3.1 创建组件	116
		8.3.2 运行项目 reactjsbook	119
	8.4	Effect Hook 的应用	119
		8.4.1 说明	119
		8.4.2 创建文件 HookExample2.js	120
		8.4.3 Effect Hook 的等价实现	121
		8.4.4 创建组件	122
		8.4.5 修改文件 index.js	124
		8.4.6 运行项目 reactjsbook	125
	8.5	其他 Hook 的应用	125
		8.5.1 useState()函数应用	125
		8.5.2 useReducer()函数应用	126
		8.5.3 useMemo()函数应用	127
		8.5.4 useRef()函数应用	128
		8.5.5 创建组件	128
		8.5.6 修改文件 index.js	129
		8.5.7 运行项目 reactjsbook	129
	习题 8		130

第 9 章 React 基础原理和高级指引 … 131

	9.1	React 基础原理	131
		9.1.1 选择性地使用 React	131
		9.1.2 JSX 表示对象	131
		9.1.3 类组件的执行顺序	133
		9.1.4 异步编程	135
		9.1.5 Fiber	135
		9.1.6 模块	136
	9.2	React 应用开发的一般步骤	136
		9.2.1 将 UI 界面分解为组件	136
		9.2.2 实现应用程序的静态版本	136
		9.2.3 确定 state	137
		9.2.4 确定 state 的放置位置	137
		9.2.5 添加反向数据流	137
	9.3	React 片段	138
		9.3.1 说明	138
		9.3.2 创建组件	138
		9.3.3 修改文件 index.js	140
		9.3.4 运行项目 reactjsbook	140
	9.4	context	141

	9.4.1 说明 141
	9.4.2 创建组件 142
	9.4.3 修改文件 index.js 144
	9.4.4 运行项目 reactjsbook 144
9.5	高阶组件 145
	9.5.1 说明 145
	9.5.2 创建组件 145
	9.5.3 修改文件 index.js 146
	9.5.4 运行项目 reactjsbook 147
9.6	ref 转发 147
	9.6.1 说明 147
	9.6.2 创建组件 148
	9.6.3 修改文件 index.js 149
	9.6.4 运行项目 reactjsbook 149
9.7	portal 149
	9.7.1 说明 149
	9.7.2 创建组件 150
	9.7.3 修改文件 index.js 151
	9.7.4 运行项目 reactjsbook 152
9.8	ref 和 DOM 152
	9.8.1 说明 152
	9.8.2 创建组件 153
	9.8.3 修改文件 index.js 156
	9.8.4 运行项目 reactjsbook 156
9.9	Web Component 157
	9.9.1 说明 157
	9.9.2 创建组件 157
	9.9.3 修改文件 index.js 158
	9.9.4 运行项目 reactjsbook 158
9.10	render props 159
	9.10.1 说明 159
	9.10.2 创建组件 159
	9.10.3 修改文件 index.js 164
	9.10.4 运行项目 reactjsbook 164
9.11	错误边界 164
	9.11.1 说明 164
	9.11.2 创建组件 165
	9.11.3 修改文件 index.js 166
	9.11.4 运行项目 reactjsbook 167

9.12 测试 167
    9.12.1 说明 167
    9.12.2 测试简单示例 167
    9.12.3 异步测试示例 169
    9.12.4 mock 测试示例 171
    9.12.5 事件测试示例 173
习题 9 174

第 10 章 React 应用开发的工具 176

10.1 包管理器 176
    10.1.1 NPM 176
    10.1.2 Yarn 177
10.2 安装 React 177
    10.2.1 CDN 链接 177
    10.2.2 Create React App 178
10.3 编译器和编辑器 178
    10.3.1 Babel 178
    10.3.2 ESLint 178
    10.3.3 Prettier 179
    10.3.4 PropTypes 179
10.4 构建工具 179
    10.4.1 webpack 179
    10.4.2 Parcel 179
10.5 服务器端渲染工具 180
    10.5.1 Next.js 180
    10.5.2 Razzle 180
    10.5.3 Gatsby 181
10.6 React Router 181
    10.6.1 说明 181
    10.6.2 创建组件 181
    10.6.3 修改文件 index.js 183
    10.6.4 运行项目 reactjsbook 183
习题 10 184

第三部分　实战篇

第 11 章 React 与 Redux 的整合开发 187

11.1 React 与 Redux 概述 187
    11.1.1 Redux 动机 187

　　　　11.1.2　Redux 核心内容 187
　　　　11.1.3　React 与 Redux 对比 189
　　11.2　计数器的开发 189
　　　　11.2.1　创建 action 189
　　　　11.2.2　创建 reducer 190
　　　　11.2.3　创建组件 190
　　　　11.2.4　修改文件 index.js 191
　　　　11.2.5　运行项目 reactjsbook 192
　　11.3　待办事项管理小工具的开发 192
　　　　11.3.1　创建 action 192
　　　　11.3.2　创建 reducer 193
　　　　11.3.3　创建组件 194
　　　　11.3.4　修改文件 index.js 198
　　　　11.3.5　运行项目 reactjsbook 198
　　习题 11 200

第 12 章　React 与 Spring Boot 的整合开发 201

　　12.1　Spring Boot 简介 201
　　　　12.1.1　Spring 的构成 201
　　　　12.1.2　Spring Boot 的特点 201
　　12.2　Spring Boot 作为后端的开发 202
　　　　12.2.1　创建项目 backendforreactjs 202
　　　　12.2.2　创建类和接口 203
　　　　12.2.3　修改后端配置文件 205
　　　　12.2.4　数据库文件 db_mediablog.sql 206
　　　　12.2.5　运行后端 Spring Boot 程序 206
　　12.3　React 作为前端的开发 206
　　　　12.3.1　修改文件 index.js 206
　　　　12.3.2　创建组件 207
　　　　12.3.3　修改前端配置文件 208
　　　　12.3.4　运行前端 React 程序 209
　　习题 12 210

第 13 章　React 与 Python 框架的整合开发 211

　　13.1　React 与 Django 的整合开发 211
　　　　13.1.1　Django 作为后端开发 211
　　　　13.1.2　运行后端 Django 程序 214
　　　　13.1.3　React 作为前端开发 215

　　　　13.1.4　运行前端 React 程序 ... 218
　13.2　React 与 Flask 的整合开发 ... 220
　　　　13.2.1　Flask 作为后端开发和运行后端 Flask 程序 ... 220
　　　　13.2.2　React 作为前端开发 ... 221
　　　　13.2.3　运行前端 React 程序 ... 221
　　　　13.2.4　打包前端 React 程序代码并手工复制到后端 ... 222
　　　　13.2.5　运行后端 Flask 程序 ... 222
　习题 13 ... 223

第 14 章　React 与 Go 的整合开发 ... 224

　14.1　Go 作为后端的开发 ... 224
　　　　14.1.1　创建项目 server 和入口文件 ... 224
　　　　14.1.2　创建 API ... 225
　　　　14.1.3　创建工具类 ... 228
　　　　14.1.4　创建数据库处理类 ... 229
　　　　14.1.5　创建数据类型 ... 231
　　　　14.1.6　数据库文件 godatabase.sql ... 232
　　　　14.1.7　运行后端 Go 程序 ... 232
　14.2　React 作为前端的开发 ... 233
　　　　14.2.1　创建项目并修改文件 index.js ... 233
　　　　14.2.2　创建用户界面 ... 233
　　　　14.2.3　创建组件 ... 234
　　　　14.2.4　运行前端 React 程序 ... 236
　习题 14 ... 237

第 15 章　案例——实现一个简易的员工信息管理系统 ... 238

　15.1　Spring Boot 作为后端的开发 ... 238
　　　　15.1.1　创建项目 excase 和实体类 ... 238
　　　　15.1.2　创建 DAO 层 ... 240
　　　　15.1.3　创建 Service 层 ... 241
　　　　15.1.4　创建 Controller 层 ... 242
　　　　15.1.5　修改后端配置文件 ... 243
　　　　15.1.6　数据库文件 studywebsite.sql ... 243
　　　　15.1.7　修改后端入口类 ... 244
　　　　15.1.8　运行后端 Spring Boot 程序 ... 244
　15.2　React 作为前端的开发 ... 244
　　　　15.2.1　修改文件 App.js 和 App.css ... 244
　　　　15.2.2　创建组件 ... 246

15.2.3　运行前端 React 程序 251
习题 15 253

附录 254

参考文献 255

第一部分　基础篇

第 1 章　React 简介与开发基础
第 2 章　React 组件
第 3 章　React 事件处理
第 4 章　React 条件渲染、列表和 key
第 5 章　React 状态管理
第 6 章　React 表单
第 7 章　React 组件的组合和继承

第 1 章

React简介与开发基础

本章先简要介绍 React 的定义、特点、发展简史，接下来结合示例分析说明 React 应用开发的基础方法，最后介绍 JSX 语法的应用开发等内容。

1.1 React 简介

1.1.1 React 的定义

React（或称为 React.js）是用于构建用户界面的 JavaScript 库，能高效、灵活地用于构建用户 UI 界面，是一种声明式编程。声明式（Declarative）编程是一种编程范式，与命令式编程相对应。命令式编程需要用算法明确地指出每一步该怎么做，而声明式编程侧重于让计算机明白要实现的目标（该做什么）。React 使创建交互式的 UI 界面变得轻而易举。无论使用什么技术栈，都可以引入 React 来开发新特性，而不需要重写现有代码。

1.1.2 React 的特点

1. 组件化开发

React 专注于提供清晰、简洁的 UI 界面解决方案。UI 界面往往是由不同的部件（如按钮、列表、标题等）构成的。这些部件可以用一些简短、独立的代码片段来实现，而这些代码片段被称为组件（Component）。React 提供了许多种不同类型的组件，开发人员也可以自定义组件。在创建组件之后，可再由这些组件构成更加复杂的 UI 界面，从而实现组件化开发。

2. 基于函数式编程的开发

从技术实现的角度来看，一个 React 组件等价于一个 JavaScript 函数。函数式编程是 React 及 React 其他生态系统中，用于实现大部分库的重要手段。在函数式编程中，函数可以是变量、数组元素、函数的参数和返回值等，并且能够像变量一样被保存、读取或在应用程序中传递。

3. 基于虚拟 DOM 的开发

当浏览器加载 HTML 文件并渲染 UI 界面时，会将构成 HTML 文件的元素（内容）变成 DOM（文档对象模型，Document Object Model）元素。一个 Web 页面（HTML 文件）往往对应一棵由 DOM 元素组成的 DOM 树。通过使用原生的 JavaScript 代码，可以直接操控 DOM，从而实现页面的更新。于是，React 把对真正 DOM 树的操作转换成对 JavaScript 树的操作，也就是对虚拟（Virtual）DOM 树的操作。React 虚拟元素是对真正 DOM 元素的描述，React 进行渲染时会将 React 虚拟元素转换成真正的 DOM 元素。虚拟元素可以分为 DOM 元素和组件元素两大类，分别对应原生 DOM 元素和自定义元素（需要将其转换成 DOM 元素）。除了特别说明外，React 元素一般是指 React 自定义元素（组件元素）。

4. 声明式编程

React 通过创建和更新虚拟元素来管理虚拟 DOM（即向虚拟 DOM 中增加虚拟元素、更新虚拟 DOM 中的元素）。开发人员在使用 React 进行应用开发时，无须逐一地写明用 DOM API 进行交互的操作命令（即命令式编程），仅指明让 React 构建什么内容（即声明式编程）即可。React 会按照声明式编程中所声明的要求，实现 UI 界面。这种以声明式编程的方式开发的 UI 界面，可以让代码更加可靠，且便于调试。

5. 基于数据更新的组件渲染

React 组件可以轻松地在应用程序中传递数据，并使状态（即数据取值）与 DOM 分离。在 React 中，数据的传输是按照自上向下的顺序单向流动的，即从父组件到子组件。这条原则让组件之间的关系变得简单且可预测。在每次更新数据后，React 会重新计算虚拟 DOM，并与上一次生成的虚拟 DOM 进行比较。当数据改变时，React 会正确、高效地更新数据和重新渲染组件，并对发生变化的部分做批量更新。

1.1.3　React 的发展简史

React 是起源于 Facebook 的内部项目，被用来开发 Instagram 网站。React 于 2013 年 5 月开源。2013 年 7 月发布了 0.3.0 版。它的声明式编程大大简化了应用程序（或简称应用）前端的开发。

2015 年 10 月发布了 0.14 版。此版本把 React 分成了 React 和 ReactDOM 两个部分。前者用于创建视图；后者负责 UI 界面的渲染工作。ReactDOM 是 React 和 DOM 之间的黏合剂。ReactDom 包括 ReactDOM.render()方法（简称为 render()方法）等功能。

2016 年 4 月发布了 15.0 版。此版本用 document.createElement()方法来代替对属性 innerHTML 的设置，让 DOM 更加轻量，在浏览器上的运行速度也较之前更快。而且，不再在文本中生成额外的<span>节点，使得 DOM 的输出结果更加简洁。

2017 年 9 月发布了 16.0 版。此版本放弃使用 React.createClass()方法来创建组件；开始支持渲染数组（此前版本只能渲染一个元素），并提供更好的处理错误的方式，更好地支持服务端渲染；是建立在 Fiber 上的全新版本。Fiber 把 React 的内核推倒重建，重写了 React

的渲染算法，提升了 React 的性能，但是几乎没有改变公开的 API，对用户使用 React 进行应用开发没有什么影响。

2019 年发布的 16.8 版本中首次使用了 Hook（钩子），为组件添加有状态的逻辑（功能）和组件之间共享（复用）逻辑提供了新的方式（不需要改动组件间结构关系）。

2020 年 10 月发布了 17.0 版。此版本主要侧重于升级、简化 React 本身。React 17.0 版还提供了一种新的升级策略（支持逐步升级）。

在 2013—2020 年的时间里，React 一直遵循 all-or-nothing 的升级策略。该策略下，要么继续使用旧版本，要么将整个应用程序升级至新版本，没有介于两者之间的情况。例如，当从 React 15.0 升级至 React 16.0 时，通常会一次升级整个应用程序。这适用于大部分应用程序的升级。但是，如果应用程序代码是在几年前编写的，并且并没有得到很好的维护，那么升级工作就会有很大的挑战性。

2021 年 3 月发布了 17.0.1 和 17.0.2 版。

2022 年 3 月发布了 18.0 版。React 18.0 版新特性包括并发渲染、过渡（transition）、自动批处理、支持 Suspense 的流式 SSR（服务端渲染）、新的严格模式（strict mode）、新的客户端和服务器渲染 API、其他新 API 等。

2022 年 4 月发布了 18.1.0 版。2022 年 6 月发布了 18.2.0 版。

1.2　React 应用开发的简单示例

1.2.1　单个 HTML 文件应用 React 的示例

一般来说，用一个文本编辑器和一个浏览器就可以完成 HTML（和 JavaScript）文件的编辑和运行。

为了前后的一致性，本书在介绍 HTML（和 JavaScript）文件的编辑和运行时，也采用集成开发环境 WebStorm。读者可以参考电子版附录 A 或其他资源来完成 WebStorm 的安装，完成安装后，双击图标就可以启动它，如图 1-1 所示（首次使用时没有项目信息）。

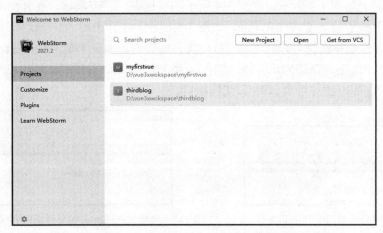

图 1-1　WebStorm 启动后的欢迎界面

在如图 1-1 所示的欢迎界面中单击 New Projcet 按钮进入项目创建界面。选择 Empty Project 类型的项目，如图 1-2 所示。创建项目时可以修改项目的位置，如图 1-2 所示中将项目所在的位置（简称为地址）修改为 d:\reactjsworkspace\firstreact；也可以采用默认地址；地址的最后一个"\"后面的内容为项目名称（如图 1-2 所示的项目名称为 firstreact）。

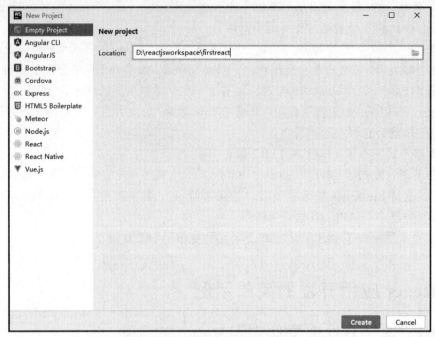

图 1-2　选择 Empty Project 类型的项目并设置项目信息的效果

创建完项目（为了区分，以 firstreact1 为项目名）后的项目目录和项目结构如图 1-3 所示，其中 firstreact1（项目名）为主要工作目录（此时目录为空），创建的 HTML 文件主要在此目录下。

在图 1-3 所示（项目名称不同，为 firstreact1）的基础上，逐步添加文件后形成的结构（折叠后的结果）如图 1-4 所示。在图 1-4 所示的项目中，右击鼠标，在弹出的菜单中选择菜单项 New 下的子菜单项 HTML File，如图 1-5 所示。在弹出的 New HTML File 对话框中输入 HTML 文件名和选择 HTML（含 XHTML）的版本，如图 1-6 所示。在输入 HTML 文件名（如 first）并选择 HTML5 file 之后，就会新增一个 HTML 文件 first.html。新增 JavaScript 文件的操作步骤和方法与新增 HTML 文件相似。

图 1-3　创建新项目后的目录和文件

图 1-4　项目 firstreact 添加文件后的目录和文件界面

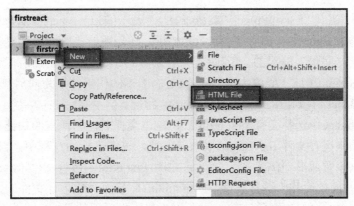

图 1-5　创建 HTML 文件的操作过程界面

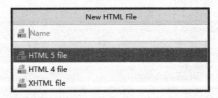

图 1-6　输入 HTML 文件名和选择文件版本的操作过程界面

在自动生成文件 first.html 的基础上,修改文件 first.html 的代码,代码如例 1-1 所示。一般来说,在创建 HTML(和 JavaScript)文件之后都需要修改文件的代码。为了叙述的简便,将创建文件并修改代码的过程简称为创建文件。

【例 1-1】　在项目 firstreact 根目录下创建文件 first.html 的代码。

```
<!DOCTYPE html>
<html lang="en">
<head>
    <meta charset="UTF-8">
    <title>React 开发示例</title>
    <!--为了在控制台中看到错误和警告信息,建议开发时使用 React 和 ReactDOM 的开发版本-- >
    <script src="https://unpkg.com/react@18.2.0/umd/react.production.min.js"></script>
    <script src="https://unpkg.com/react-dom@18.2.0/umd/react-dom.production.min.js"></script>
    <script src="https://unpkg.com/@babel/standalone@7.18.0/babel.min.js"></script>
</head>
<body>
<div id="root"><!--页面内容会被渲染到此处,用 React 显示内容的位置--></div>
<script type="text/babel">
    ReactDOM.createRoot(document.getElementById('root'))   //此处 root 和 div 中的 root 要一致
        .render(<h1>Hello, world!</h1>);                    //要渲染的内容
</script>
</body>
</html>
```

在例 1-1 中，语句"<script type="text/babel">"表示<script type="text/babel">和</script>之间的脚本代码由 Babel.js（简称为 Babel）解析，还表明此段脚本代码是用 React 的 JSX（JavaScript 的语法扩展）编写的。Babel 是一个 JavaScript 编译器（或称为工具链），主要用于将 ECMAScript 2015（或称为 ES 6）及更高版本的代码转换为向后兼容的 JavaScript 版本，借助于 Babel 可以使用最新的 JavaScript 特性。

在项目 firstreact 中选择文件 first.html 后，右击文件名，在弹出的快捷菜单中选择 Run 'first.html'菜单项来运行文件 first.html，如图 1-7 所示。在运行的过程中，自动打开 Chrome 浏览器（也可以设置成其他浏览器）并访问文件 first.html，运行文件 first.html 的效果如图 1-8 所示。这些操作简称为运行文件（如 first.html），后面章节运行文件的方法与此相同。注意，在运行文件时要保证网络连接正常，因为代码中采用的是内容分发网络（CDN）方式来访问网络中的 React 等内容的。

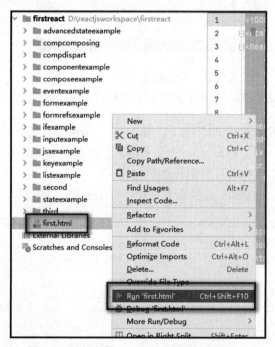

图 1-7 运行文件 first.html 的操作过程界面

图 1-8 运行文件 first.html 的效果

1.2.2 元素渲染说明

在创建 React 元素后，ReactDOM 就提供了在浏览器中渲染 React 元素（包括它的子元素）所需要的方法。假设 HTML 文件（如例 1-1）有一个<div>，代码为<div id="root">…

</div>，将其称为 DOM 根节点，该节点内的所有内容都将由 ReactDOM 管理。元素是构成 React 应用程序的最小单元。要将一个 React 元素（如 element）渲染到 DOM 根节点（如 root）中，代码如例 1-2 所示。

【例 1-2】 将一个 React 元素（如 element）渲染到 DOM 根节点的代码。

```
const element = <h1>Hello, world</h1>;
ReactDOM.createRoot(document.getElementById('root')).render(element);
```

由于 React 元素是不可变对象，因此一旦被创建，就无法更改它的子元素或者属性。更新 UI 界面的唯一方法是创建一个全新的元素，并将其传入 render()方法中进行渲染。

React 认为，渲染逻辑在本质上与其他 UI 界面逻辑是内在耦合的。例如，在 UI 界面中需要绑定处理事件，在某些时刻，状态发生变化时需要通知到 UI 界面，需要在 UI 界面中展示准备好的数据。

在渲染所有输入内容之前，ReactDOM 会进行转义，即所有的内容在渲染之前都被转换成字符串。这可以确保在应用程序中不会注入那些并非开发人员编写的内容。这样可以有效地防止 XSS（Cross-Site Scripting，跨站脚本，为避免和层叠样式表的缩写混淆，故缩写为 XSS）的攻击。

视频讲解

1.2.3 两个文件应用 React 的示例

在 firstreact 根目录（即 firstreact）下创建 second 子目录，在 firstreact\second 目录下创建文件 second.html，代码如例 1-3 所示。

【例 1-3】 在 firstreact/second 目录下创建文件 second.html 的代码。

```
<!DOCTYPE html>
<html lang="en">
<head>
    <meta charset="UTF-8">
    <title>React 开发示例</title>
    <script src="https://unpkg.com/react@18.2.0/umd/react.production.min.js"></script>
    <script src="https://unpkg.com/react-dom@18.2.0/umd/react-dom.production.min.js"></script>
    <script src="https://unpkg.com/@babel/standalone@7.18.0/babel.min.js"></script>
</head>
<body>
<div id="root"></div>
<!--将 js 或 jsx 代码作为独立文件，并引入-->
<script type="text/babel" src="index.js"></script>
</body>
</html>
```

在例 1-3 中的<body> 标签内的任意位置放置一个容器<div>。给这个 <div> 加上唯一的 id 属性（root），从而使 JavaScript 代码可以找到它，并在其中渲染 React 组件。根据需要,可以在一个页面上放置多个独立的 DOM 容器。它们通常是空标签,React 会替换 DOM 容器内任何已有内容。

在 firstreact\second 目录下创建文件 index.js,代码如例 1-4 所示。

【例 1-4】 在 firstreact\second 目录下创建的文件 index.js 的代码。

```
//React 渲染方法的调用
ReactDOM.createRoot(document.getElementById('root'))//要和HTML文件中的root对应
    .render(<h1>Hello, world!</h1>);              //要渲染的内容
```

运行文件 second.html,效果如图 1-9 所示。

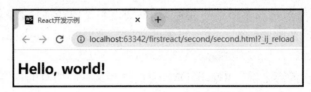

图 1-9　运行文件 second.html 的效果

在项目 firstreact 根目录下创建 third 子目录,在 firstreact\third 目录下创建文件 third.html,third.html 代码与例 1-3 所示的文件 second.html 代码相同(只是 HTML 文件名不同)。在 firstreact\third 目录下创建文件 index.js,代码如例 1-5 所示。

【例 1-5】 在 firstreact\third 目录下创建的文件 index.js 的代码。

```
const divReact = document.getElementById('root');//获得内容的挂载(渲染)处
//JSX 语法表示基础信息、数据
const infoMap={
    helloMessage:'Hello, world!',
}
//JSX 语法表示要渲染的内容
const reactSpan = (
      <h1>{infoMap.helloMessage}</h1>
);
ReactDOM.createRoot(divReact).render(reactSpan);//将内容渲染到HTML中对应的位置
```

运行文件 third.html,效果如图 1-10 所示。

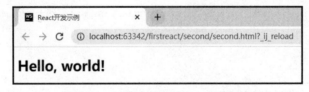

图 1-10　运行文件 third.html 的效果

1.2.4　应用 React 的示例对比分析

React 并没有采用将标签与逻辑分离到不同文件这种方式,而是将二者都存放在称之为 "组件" 的松耦合单元之中来实现关注点分离。

结合图 1-8~图 1-10，可以发现 1.2.1 节的 first.html 与 1.2.3 节的 second.html、third.html 输出结果相同，三种实现方法的功能相同。1.2.3 节的两个示例的开发都是通过两个文件实现（一个是 HTML 文件，另一个是文件 index.js）的，将 first.html 文件内容进行了分离；其中 second.html 和 third.html 两个 HTML 文件代码相同（在后面的章节中，除非特别说明，HTML 文件的代码一般也相同），不同的是两个 index.js 文件。firstreact\third 目录下文件 index.js 是对 firstreact\second 目录下创建文件 index.js 的进一步分解。这些分解是为了更好地理解 React 开发，避免代码的重复。后面章节在介绍 React 的基础应用开发时，以 HTML 文件和类似 firstreact\third 目录下的文件 index.js 两个文件组合的方式为主要开发方式。

1.3 JSX 应用开发入门

视频讲解

1.3.1 JSX 说明

React 官方文档建议在 React 应用开发时使用 JSX。JSX 是对 JavaScript 语法的扩展，它具有 JavaScript 的全部功能，还提供了在 JavaScript 代码中直接使用 HTML 标签的语法糖。

JSX 可以很好地描述 UI 界面应该呈现的本质形式。React 不强制要求使用 JSX，但将 JSX 和 UI 界面放在一起时，会在视觉上有辅助作用。它还可以使 React 应用程序的代码中显示更多有用的错误和警告消息。

在 JSX 中可以任意使用 JavaScript 表达式，只需要用一个大括号把表达式括起来。React 元素实际上都是一个 JavaScript 对象，可以在应用时把它保存在变量中或者作为参数传递。JSX 可以生成 React 元素。

在编译之后，JSX 表达式会被转为普通 JavaScript 函数调用，并且对其取值后得到 JavaScript 对象。JSX 有算术、条件、嵌入、对象、函数、增强函数、数组、样式、注释等表达式。为了便于阅读，可将 JSX 拆分为多行。

1.3.2 JSX 综合应用示例

在项目 firstreact 根目录下创建 jsxexample 子目录，在 firstreact\jsxexample 目录下创建文件 jsxexample.html，代码与例 1-3 所示的代码相同。在 firstreact\jsxexample 目录下创建文件 index.js，代码如例 1-6 所示。

【例 1-6】 在 firstreact\jsxexample 目录下创建的文件 index.js 的代码。

```
//文件第 1 部分——指明要挂载（渲染）的 div 节点
const divReact = document.getElementById('root');
//文件第 2 部分——提供基础和公共管理的数据信息
//infoMap 中的内容均为字符串形式，用于输出提示信息
const infoMap= {
    nameZsf:'张三丰',
    ageZsf:88,
    nameLs:'李斯',
    ageLs:50,
```

```
        ageUnit:'岁',
        font20:20,
        bold:'bold',
        red:'red',
        font12:12,
        italic:'italic',
        blue:'blue',
        noUserInfo:'没有用户信息',
        titleInfo:'JSX 简单示例',
        textExample:'直接显示文本示例：Hello, world!',
        expressionInput:'表达式示例：使用大括号{}来表示表达式，例如{3+6}=',
        conditionInput:'条件表达式示例：例如{1 == 1 ? "true" : "false"}的返回结果为',
        objectInput:'对象表达式示例：例如"{objectUser.name} {objectUser.age}岁"的
返回结果为',
        functionInput:'函数表达式示例：例如"{formatUserinfo(objectUser)}"的返回结
果为"',
        functionWithArgsInput:'有参数的增强函数表达式示例：例如"{chooseUserinfo
(objectUser)}"的返回结果为"',
        functionNoArgsInput:'无参数的增强函数表达式示例：例如"{chooseUserinfo()}"的
返回结果为"',
        arrayInput:'数组表达式示例：例如"{arrayUserinfo}"的返回结果为"',
        jsxContentForStyle1:'样式表达式示例：此处为style1，请注意style1和style2字
体的大小区别',
        jsxContentForStyle2:'样式表达式示例：此处为style2，请注意style1和style2字
体的大小区别',
        correctCommentInput:'注释表达式示例：注释的外部也要用大括号，即注释也是一种表达
式；例如"{/* a+b=1 */}"为正确注释方式',
        errorCommentInput:'而/* a+b=1 */为错误注释方式'
    }
//文件第3部分——自定义结构、函数（组件）等
const objectUser={
    name:infoMap.nameZsf,
    age:infoMap.ageZsf,
}
var arrayUserinfo = [
    <span>{infoMap.nameLs}</span>,
    <span>{infoMap.ageLs}</span>,
    <span>{infoMap.ageUnit}</span>
];
const css_p_lg = {
    fontSize: infoMap.font20,
    fontStyle: infoMap.bold,
    color: infoMap.red
};
const css_p_sm = {
    fontSize: infoMap.font12,
    fontStyle: infoMap.italic,
    color: infoMap.blue
};
```

```
        function formatUserinfo(userinfo) {
            return userinfo.name + " " + userinfo.age + infoMap.ageUnit;
        }
        function chooseUserinfo(userinfo) {
            if(userinfo) {
                return userinfo.name + " " + userinfo.age + infoMap.ageUnit;
            } else {
                return infoMap.noUserInfo;
            }
        }
        //文件第 4 部分——指明要渲染的内容
        const exampleJSX = (
            <span>
                {/*请注意像此处标签内的 JSX 代码里编写注释时必须采用"{、}"方式*/}
                {/*请将此处代码和输出结果对照,加强对 JSX 语法的学习*/}
                <h2 align="center">{infoMap.titleInfo}</h2>
                {/*由于 exampleMap 中的内容均为字符串,外面加上大括号只是直接显示字符串内容*/}
                <h3>{infoMap.textExample}</h3>
                <hr/>
                {/*对于大括号中的表达式,按照对应语法进行处理,例如 3+6=9,用来输出表达式计算结果*/}
                <h3>{infoMap.expressionInput}{3+6}</h3>
                <hr/>
                <h3>{infoMap.conditionInput}{1 == 1 ? "true" : "false"}</h3>
                <hr/>
                <h3>{infoMap.objectInput}{'"'}{objectUser.name}{objectUser.age}
{'岁"'}</h3>
                <hr/>
                <h3>{infoMap.functionInput}{formatUserinfo(objectUser)}{'"'}</h3>
                <hr/>
                <h3>{infoMap.functionWithArgsInput}{chooseUserinfo(objectUser)}
{'"'}</h3>
                <hr/>
                <h3>{infoMap.functionNoArgsInput}{chooseUserinfo()}{'"'}</h3>
                <hr/>
                <h3>{infoMap.arrayInput}{arrayUserinfo}{'"'}</h3>
                <hr/>
                <h3 style={css_p_lg}>{infoMap.jsxContentForStyle1}</h3>
                <hr/>
                <h3 style={css_p_sm}>{infoMap.jsxContentForStyle2}</h3>
                <hr/>
                <h3>{infoMap.correctCommentInput}{/* This is a right JSX Comment
Expression. */}</h3>
                <h3>{infoMap.errorCommentInput}</h3>
            </span>
        );
        //文件第 5 部分——渲染语句
        ReactDOM.createRoot(divReact).render(exampleJSX);//将内容渲染到 HTML 中对应的位置
```

如例 1-6 所示,可以将 index.js 文件分成 5 个主要部分:第 1 部分是辅助工作,如指明要挂载（渲染）的 div 节点、导入模块（文件）；第 2 部分提供基础和公共管理的数据信息；第 3 部分是自定义结构、函数（组件）等内容；第 4 部分指明要渲染的内容；第 5 部分是渲染语句。

1.3.3 JSX 综合运行效果

运行文件 jsxexample.html，效果如图 1-11 所示。

图 1-11　运行文件 jsxexample.html 的效果

习题 1

一、简答题

1. 简述对 React 的理解。
2. 简述对 JSX 的理解。

二、实验题

1. 完成单个 HTML 文件应用 React 的示例。
2. 完成多个文件应用 React 的示例。
3. 完成 JSX 的开发示例。

第 2 章

React 组 件

本章先简要介绍组件，再介绍函数组件和类组件的应用开发、组件参数和组合组件、组件的分解和组合、组件的生命周期等内容。

2.1 React 组件概述

视频讲解

2.1.1 组件和自定义组件

可以将应用程序的 UI 界面拆分成独立可复用的代码片段（即组件），并对每个组件进行开发。组件名称必须以大写字母开头。React 会将以小写字母开头的组件视为原生 DOM 标签。例如，<div/>代表 HTML 标签 div，而<Welcome/> 则代表组件 Welcome。

组件从概念上类似于 JavaScript 函数（function）。它接收任意的入参（即 props），并返回用于描述页面显示内容的 React 元素。React 元素可以是 DOM 标签（元素），也可以是用户自定义的组件。当 React 元素为用户自定义组件时,它会将 JSX 所接收的属性(attributes)以及子组件（children）转换为单个对象传递给组件，这个对象被称为 props。

以通过自定义组件在页面上渲染显示 "Hello, React Component!" 为例。先在项目目录 firstreact 下创建 exhellocomponent 子目录，在 firstreact\exhellocomponent 目录下创建文件 exhellocomponent.html，代码与例 1-3 所示的代码相同。在 firstreact\exhellocomponent 目录下创建文件 index.js，代码如例 2-1 所示。

【例 2-1】 在 firstreact\exhellocomponent 目录下创建的文件 index.js 的代码。

```
const divReact = document.getElementById('root');
function Welcome(props) {
    return <h1>Hello, {props.name}</h1>;
}
const helloComp = (<Welcome name="React Component!" />);
```

```
ReactDOM.createRoot(divReact).render(helloComp);
```

运行文件 exhellocomponent.html，效果如图 2-1 所示。

图 2-1　运行文件 exhellocomponent.html 的效果

例 2-1 的执行步骤包括：调用 ReactDOM.createRoot()方法得到一个 ReactDOMRoot 实例后，调用 render()方法，并传入<Welcome name= "React Component!" /> 作为参数；React 调用 Welcome 组件，并将{name: 'React Component!'}作为参数传入；Welcome 组件将<h1>Hello, React Component!</h1>元素作为返回值；React DOM 将 DOM 高效地更新（渲染）为<h1>Hello, React Component!</h1>。

2.1.2　函数组件和类组件

定义组件最简单的方式就是编写 JavaScript 函数，这类组件被称为函数组件。例如，例 2-1 中的 Welcome 组件就是函数组件。还可以使用类（class）来定义组件，这类组件被称为类组件。

定义类组件时，类组件需要继承类 React.Component。还需在类组件中实现 render()方法，render()方法是类组件中唯一必须实现的方法。当它被调用时，它会检查当前的 props 和状态（state）的变化情况并返回 React 元素、数组或 fragments（可以返回多个元素）、字符串或数值类型（在 DOM 中会被渲染为文本节点）、portal（可以渲染子节点到不同的 DOM 子树中）、布尔类型或 null（什么都不渲染）。state 与 props 类似，但是 state 是私有的，并且完全受控于所在组件。

将函数组件转成类组件的步骤：创建一个同名的继承类 React.Component 的类，添加一个空的 render()方法，将函数体移动到 render()方法之中，在 render()方法中使用 this.props 替换 props，删除剩余的空函数声明。ES6 为 JavaScript 添加类的语法后，React 官方推荐使用类组件；目前 React 官方推荐使用函数组件（类组件也可以正常使用）。在不久的将来，类组件可能会被弃用。

2.2　函数组件和类组件的应用开发

2.2.1　开发示例

在 firstreact 根目录下创建 componentexample 子目录，在 firstreact\componentexample 目录下创建文件 componentexample.html，代码与例 1-3 所示的代码相同。在 firstreact\

componentexample 目录下创建文件 index.js，代码如例 2-2 所示。

【例 2-2】 在 firstreact\ componentexample 目录下创建的文件 index.js 的代码。

```
const divReact = document.getElementById('root');
const infoMap= {
    functionComInfo:'Function Component',
    classComInfo:'Class Component',
    titleInfo:'函数组件和类组件的简单示例',
    functionCompOutput:'函数组件的示例: function compName() {\n' +
        '    return <h3>Function Component</h3>;\n' +
        '}结果为',
    classCompOutput:'类组件的示例: class FullReactComp extends React.Component {\n' +
        '    render() {\n' +
        '        return <h3>Class Component</h3>;\n' +
        '    }\n' +
        '}结果为'
}
//定义函数组件，组件名大写
function FunctionReactComp() {
    return <h3>{infoMap.functionComInfo}</h3>;
}
//定义类组件
class ClassReactComp extends React.Component {
    render() {
        return <h3>{infoMap.classComInfo}</h3>;
    }
}
//声明变量
const functionComp =<FunctionReactComp/> ;
const classComp = <ClassReactComp/>;
//要渲染的内容
const exampleComp = (
    <span>
        <h2 align="center">{infoMap.titleInfo}</h2>
        <hr/>
        <h3>{infoMap.functionCompOutput}</h3>
        {functionComp}
        <hr/>
        <h3>{infoMap.classCompOutput}</h3>
        {classComp}
    </span>
);
ReactDOM.createRoot(divReact).render(exampleComp);
```

2.2.2 运行效果

运行文件 componentexample.html，效果如图 2-2 所示。

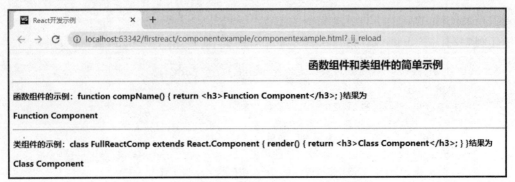

图 2-2　运行文件 componentexample.html 的效果

2.3　组件参数和组合组件

2.3.1　说明

在 React 应用程序中，按钮、表单、对话框，甚至整个 UI 界面（屏幕）的内容通常都以组件的形式表示。组件可以在其输出中引用其他组件。于是，可以用同一组件与其他组件组合来抽象出任意层次的细节。将 React 集成到现有的应用程序中，可能需要使用像 Button 这样的小组件，并自下而上地将这类组件逐步应用到更多地方。一般来说，每个新的 React 应用程序的顶层组件都是 App 组件。

组件不能修改传给自身的参数 props。this.props 包括调用本组件的组件定义的 props。注意，this.props.children 是一个特殊的 props，通常由 JSX 表达式中的子组件组成，而非组件自身定义。defaultProps 可以为类组件添加默认 props。这一般用于 props 未被赋值，但又不能为 null 的情况。

在不修改组件 state 的情况下，每次调用类组件 render()方法时都返回相同的结果，并且它不会直接与浏览器交互。如果需要与浏览器进行交互，可以在 componentDidMount()方法或其他生命周期方法中执行交互操作。如果 shouldComponentUpdate()方法返回 false，就不会调用 render()方法。

2.3.2　开发示例

在项目 firstreact 根目录下创建 compcomposing 子目录，在 firstreact\compcomposing 目录下创建文件 compcomposing.html，代码与例 1-3 所示的代码相同。在 firstreact\compcomposing 目录下创建文件 index.js，代码如例 2-3 所示。

【例 2-3】　在 firstreact\compcomposing 目录下创建的文件 index.js 的代码。

```
const divReact = document.getElementById('root');
const infoMap= {
    formTitle:'用户登录',
    userName:'用户名: ',
    passwd:'密码(*):',
```

```
    submitValue:'登录',
    args1begin:'传的参数props是"',
    args1end:'"。',
    args2:'没有传参。',
    nameLs:'李斯',
    defaultInfo:'使用this传入的默认（缺省）参数值为"',
    nameWym:'王阳明',
    nameZsf:'张三丰',
    part1title:'组件组合的简单示例',
    part2title:'组件传参props的简单示例',
    part3title:'组件只读参数（默认值，无法修改）的简单示例',
    part4title:'类组件参数默认（缺省）值的简单示例',
}
//无参组件
function FormTitle() {
    return <h4>{infoMap.formTitle}</h4>;
}
function InputComp(){
    return <input type="text"/>;
}
function UserName() {
    const userName = (
        <span><div>{infoMap.userName}</div><InputComp/></span>
    );
    return userName;
}
function Password() {
    const passwd = (
        <span><div>{infoMap.passwd}</div><InputComp/></span>
    );
    return passwd;
}
function Submit() {
    const submit = (
        <p><button>{infoMap.submitValue}</button></p>
    );
    return submit;
}
//有参组件
function PropsReactComp(props) {
    if(props) {
        return <p>{infoMap.args1begin}{props.name}{infoMap.args1end}</p>;
    } else {
        return <p>{infoMap.args2}</p>;
    }
}
//无参组件的组合
function FormLogin() {
    return (
```

```jsx
            <div id="id-div-formlogin">
                <FormTitle/>
                <UserName/>
                <Password/>
                <Submit/>
            </div>
        );
    }
    function UserName2(props) {
        const userName = (
            <span>
                <div>{infoMap.userName}</div>
                <input type="text" value={props.userName} />
            </span>
        );
        return userName
    }
    //有参组件的组合
    function FormLogin2() {
        return (
            <div id="id-div-formlogin2">
                <FormTitle/>
                <UserName2 userName={infoMap.nameLs}/>
                <Password/>
                <Submit/>
            </div>
        );
    }
    //使用默认值作为参数
    //使用关键字this
    class PropsReactComp2 extends React.Component {
        render() {
            return <p>{infoMap.defaultInfo}{this.props.default}{infoMap.args1end}</p>;
        }
    }
    PropsReactComp2.defaultProps = {
        default: infoMap.nameWym
    };
    //声明创建一个无参的组合组件实例
    const formLogin = <FormLogin/>;
    //声明创建一个组件实例,并传递参数
    const eleProps = <PropsReactComp name={infoMap.nameZsf}/>;
    //声明创建一个有参的组合组件实例,并传递参数
    const formLogin2 = <FormLogin2/>;
    const eleProps2 = <PropsReactComp2/>;
    const exampleComp = (
        <span>
            <h3 align="center">{infoMap.part1title}</h3>
```

```
        {formLogin}
        <hr/>
        <h3 align="center">{infoMap.part2title}</h3>
        {eleProps}
        <hr/>
        <h3 align="center">{infoMap.part3title}</h3>
        {formLogin2}
        <hr/>
        <h3 align="center">{infoMap.part4title}</h3>
        {eleProps2}
    </span>
);
ReactDOM.createRoot(divReact).render(exampleComp);
```

2.3.3 运行效果

运行文件 compcomposing.html，效果如图 2-3 所示。

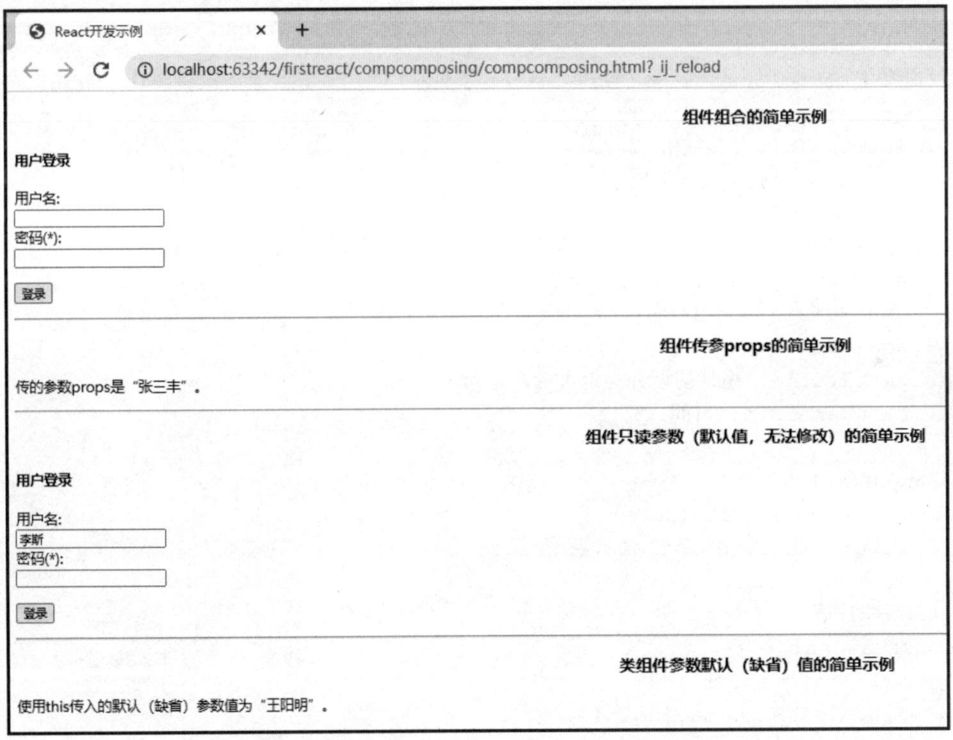

图 2-3 运行文件 compcomposing.html 的效果

2.4 组件的分解和组合

视频讲解

2.4.1 说明

为了更好地对开发工作进行分工，以及便于代码的编写、复用、阅读和后期维护，在进行组件开发时要对组件进行分解。将一个组件拆分成多个更小的组件，即提取组件。再在提取组件的基础上，对组件进行组合，实现相关的功能。采用声明式编程风格提取组件并把小型组件组合在一起，这样编写出来的代码简洁明了，更易于阅读。

虽然，提取组件是一项繁重的工作，但是在大型应用程序中构建可复用组件库是完全值得的。根据经验来看，如果 UI 界面中有一部分被多次使用（如 Button），或者组件足够复杂，就可以将重复代码提取成一个组件，以便于代码的复用。

2.4.2 开发示例

在 firstreact 根目录下创建 compdispart 子目录，在 firstreact\compdispart 目录下创建文件 compdispart.html，代码与例 1-3 所示的代码相同。在 firstreact\compdispart 目录下创建文件 index.js，代码如例 2-4 所示。还需要在 firstreact\compdispart 目录下创建 images 子目录，并且在 firstreact\compdispart\images 目录下准备图片文件 yanhui.jpg。

【例 2-4】 在 firstreact\compdispart 目录下创建的文件 index.js 的代码。

```js
const divReact = document.getElementById('root');
const infoMap= {
//需要在 compdispart 目录下创建 images 子目录，并且在该目录下准备图片 yanhui.jpg
    url:'images/yanhui.jpg',
    loadingInfo:'装载中...',
    nickname:'孔子的学生',
    userName:'颜回',
    age:'38',
    nicknameInfo:'昵称：',
    userNameInfo:'用户名：',
    ageInfo:'年龄：',
    part1title:'组件的切分与组合的简单示例',
    dateNameInfo:'时间：',
}
//格式化时间
function formatDate(date) {
    return date.toLocaleDateString();
}
//封装图片信息
const avatar = {
    url: infoMap.url,
    alt: infoMap.loadingInfo,
    nickname:infoMap.nickname
};
//封装用户信息
const userinfo = {
    name: infoMap.userName,
    age: infoMap.age,
};
//封装时间信息
const date = {
```

```
        date: formatDate(new Date())
    };
    //图片信息组件
    function Avatar(props) {
        return (
            <span className="cssAvatar">
                <img className="cssAvaterImg"
                    src={props.avatar.url}
                    alt={props.avatar.alt}
                />
                <p className="p-center">{infoMap.nicknameInfo}{props.avatar.nickname}</p>
            </span>
        );
    }
    //用户信息组件
    function UserInfo(props) {
        return (
            <span className="cssUserinfo">
                <p className="p-left">{infoMap.userNameInfo}{props.userinfo.name}</p>
                <p className="p-left">{infoMap.ageInfo}{props.userinfo.age}</p>
            </span>
        );
    }
    //用户详细信息组件
    function UserDetail(props) {
        return (
            <div className="cssUserDetail">
                <Avatar avatar={props.avatar} />
                <hr/>
                <UserInfo userinfo={props.userinfo} />
                <hr/>
                <span className="cssDate">
                    <p className="p-right">{infoMap.dateNameInfo}{props.curdate.date}</p>
                </span>
            </div>
        );
    }
    //用户详细信息组件实例，展示用户信息中的细节数据
    const userDetail = <UserDetail
        avatar={avatar}
        userinfo={userinfo}
        curdate={date}
        />;
    const exampleComp = (
        <span>
```

```
            <h3 align="center">{infoMap.part1title}</h3>
            {userDetail}
        </span>
);
ReactDOM.createRoot(divReact).render(exampleComp);
```

2.4.3 运行效果

运行文件 compdispart.html，效果如图 2-4 所示。注意图 2-4 中的时间为运行文件时的时间。

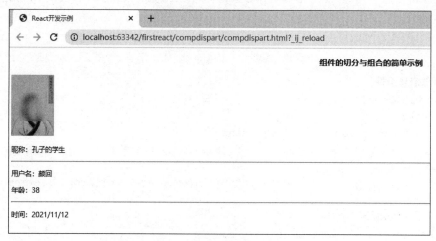

图 2-4 运行文件 compdispart.html 的效果

2.5 组件的生命周期

2.5.1 概述

组件第一次被渲染到 DOM 中的过程被称为挂载（mount）。删除 DOM 中组件的过程被称为卸载（unmount）。组件从挂载到卸载及其他环节（过程）称为组件的生命周期。可以为组件声明一些特殊的方法，在挂载、卸载组件或组件生命周期中其他时间节点让组件去执行这些方法。这些方法叫作生命周期方法。生命周期方法由 React 主动调用。每个组件都包含一些默认生命周期方法，也可以重写这些方法，以便在组件的执行过程中执行这些方法。

组件挂载时生命周期方法的调用顺序为：constructor()方法（构造方法）、getDerivedStateFromProps()方法、render()方法、componentDidMount()方法。当组件的参数或状态发生变化时会触发组件进行更新。组件更新时生命周期方法调用顺序为：getDerivedStateFromProps() 方法、shouldComponentUpdate() 方法、render() 方法、getSnapshotBeforeUpdate()方法、componentDidUpdate()方法。组件卸载时会调用 componentWillUnmount()方法。在渲染过程、生命周期或子组件的构造方法（或称为构造函数）中抛出错误时，会调用 getDerivedStateFromError()方法和 componentDidCatch()方法。

本章将介绍 constructor()方法、componentDidMount()方法、componentDidUpdate()方法、componentWillUnmount()方法等常用方法，对于其他不常用的方法请读者参考官方文档的说明。

2.5.2　constructor()方法

如果不需要初始化 state 和进行方法绑定，就不需要为组件实现构造方法 constructor(props)。构造方法 constructor(props)可简称为 constructor()方法。

React 组件挂载前会调用它的构造方法。在实现构造方法时，应在其他语句之前调用父类 React.Component 的 super(props)。构造方法通常用于初始化内部 state 或进行事件处理两种情况。

如果组件需要使用内部 state，可直接在构造方法中将 this.state 赋值初始化成 state 即可。要避免将 props 的值赋值给 state。如果需要在其他方法中为 state 赋值，应使用 setState()方法。

2.5.3　componentDidMount()方法

componentDidMount()方法会在组件挂载后被立即调用。依赖于 DOM 节点的初始化代码应该放在此方法中。例如，需要通过网络请求获取数据时，实例化请求的代码就可以放在此方法中。添加订阅功能的代码也适合放在此方法中。如果在此方法中添加了订阅功能，就要在 componentWillUnmount()方法中取消订阅。

可以在 componentDidMount()方法中调用 setState()方法。它将触发额外渲染，此渲染会发生在浏览器更新 UI 界面之前。如果渲染依赖于其大小或位置的 DOM 节点时，可以使用此方式处理。但须谨慎使用此种编码方式，因为它会导致性能问题。

2.5.4　componentDidUpdate()方法

componentDidUpdate(prevProps, prevState, snapshot)方法可以简称为 componentDidUpdate()方法。componentDidUpdate()方法会在组件更新后被立即调用。首次渲染不会执行此方法。组件更新后，可以在此方法中对 DOM 进行操作。如果需要对更新前后的 props 进行比较，也可以选择在此方法中进行。

可以在 componentDidUpdate()方法中直接调用 setState()方法。注意，setState()方法必须被包裹在一个条件语句里；否则，不仅会导致死循环，还会导致额外的重新渲染，影响组件性能。

如果组件实现了 getSnapshotBeforeUpdate()方法，那么它的返回值将作为 componentDidUpdate()的第三个参数 snapshot 传入。否则，componentDidUpdate()的第三个参数为 undefined。如果 shouldComponentUpdate()方法的返回值为 false，就不会调用 componentDidUpdate()方法。

2.5.5　componentWillUnmount()方法

componentWillUnmount()方法会在组件卸载及销毁之前直接调用。可以在此方法中执行必要的清理操作。例如，清除 timer、取消网络请求或清除在 componentDidMount()方法中添加的订阅等。

注意，不要在 componentWillUnmount()方法中调用 setState()方法，因为此时的组件将永远不会重新渲染，且组件被卸载后将永远无法再挂载它。

2.5.6 开发示例

在 firstreact 根目录下创建 exoflifecycle 子目录，在 firstreact\exoflifecycle 目录下创建文件 exoflifecycle.html，代码与例 1-3 所示的代码相同。在 firstreact\exoflifecycle 目录下创建文件 index.js，代码如例 2-5 所示。

【例 2-5】 在 firstreact\exoflifecycle 目录下创建的文件 index.js 的代码。

```
const divReact = document.getElementById('root');
const infoMap={
    constructorInfo:'构造函数时输出的结果',
    mountInfo:'加载时输出的结果',
    unMountInfo:'卸载时输出的结果',
    tickInfo:'执行 tick()方法时输出的结果',
    title1Info: '执行步骤：',
    dialogInfo:'弹出对话框，显示',
    tickInfo2:'调用 tick 方法',
    colon:': ',
    fullStop:'。'
}
class Counter extends React.Component {
    constructor(props) {
        super(props);
        this.state = {
            startNumber:props.startnumber,
            step:props.step,
            multi:2,
            resultNumber:0,
        };
        this.state.resultNumber=this.state.step;
        alert(infoMap.constructorInfo+this.state.resultNumber);//结果取值为 2
    }
    componentDidMount() {
        this.state.resultNumber+=this.state.startNumber*this.state.step*this.state.multi;
        alert(infoMap.mountInfo+this.state.resultNumber)    //结果取值为 60
    }
    componentWillUnmount() {
        this.state.resultNumber -= this.state.startNumber*this.state.step;
        alert(infoMap.unMountInfo+this.state.resultNumber)
    }
    tick() {
        this.state.resultNumber=20;
```

```
            alert(infoMap.tickInfo+this.state.resultNumber)   //结果取值为 20
    }
    render() {
        return (
            <div>
                <h2>{infoMap.title1info}</h2>
    <h3>{infoMap.dialogInfo}{infoMap.constructorInfo}{infoMap.colon}{this.
state.resultNumber}</h3>
                <h3>{infoMap.tickInfo2} {this.tick()}{infoMap.fullStop}</h3>
    <h3>{infoMap.dialogInfo}{infoMap.tickInfo}{infoMap.colon}{this.state.
startNumber*2}</h3>
    <h3>{infoMap.dialogInfo}{infoMap.tickInfo}{infoMap.colon}{this.state.
startNumber*2}</h3>
    <h3>{infoMap.dialogInfo}{infoMap.mountInfo}{infoMap.colon}{this.state.
startNumber*6}</h3>
                <h3>{infoMap.tickInfo2}{this.tick()}{infoMap.fullStop}</h3>
            </div>
        );
    }
}
const lifeComp = (<Counter startnumber={10} step={2}></Counter>);
ReactDOM.createRoot(divReact).render(lifeComp);
```

2.5.7 运行效果

运行文件 exoflifecycle.html，效果如图 2-5 所示。单击图 2-5 所示的对话框中的"确定"按钮，效果如图 2-6 所示。单击图 2-6 所示的对话框中的"确定"按钮，效果如图 2-6 所示（两次弹出的对话框及其内容相同）。再次单击图 2-6 所示的对话框中的"确定"按钮，效果如图 2-7 所示。单击图 2-7 所示的对话框中的"确定"按钮，效果如图 2-8 所示。

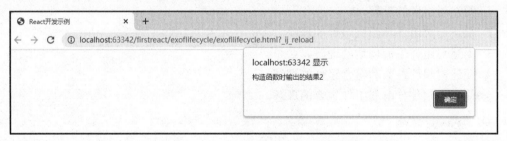

图 2-5　运行文件 exoflifecycle.html 的效果

图 2-6　单击图 2-5 所示的对话框中的"确定"按钮的效果

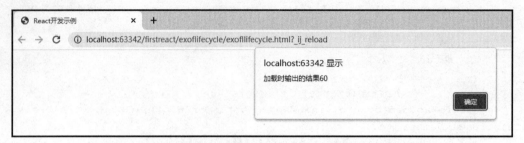

图 2-7　再次单击图 2-6 所示的对话框中的"确定"按钮的效果

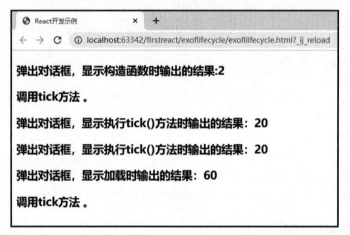

图 2-8　单击图 2-7 所示的对话框中的"确定"按钮的效果

习题 2

一、简答题
1. 简述对组件的理解。
2. 简述对类组件和函数组件的理解。
3. 简述对组件生命周期的理解。
4. 简述对组件参数和组合组件的理解。
5. 简述对组件分解和组件组合的理解。

二、实验题
1. 完成类组件的开发示例。
2. 完成函数组件的开发示例。
3. 完成组件参数的开发示例。
4. 完成组件参数和组合组件的开发示例。
5. 完成组件分解和组件组合的开发示例。
6. 完成组件生命周期的开发示例。

第 3 章

React 事件处理

本章介绍 React 事件处理概述、鼠标事件处理、焦点事件处理、键盘事件处理和图像事件处理的应用开发。

3.1 React 事件处理概述

3.1.1 事件

JavaScript 与 HTML 的交互是通过事件（Event）实现的。事件是指用户或浏览器执行的某种动作，如用户单击鼠标等。事件代表文档（Document）或浏览器窗口（Window）等某个有意义的运行时刻。

事件概念最早是在 IE3 和 Navigator2 中出现的，当时的用意是把某些表单处理工作从服务器转移到浏览器。DOM2 开始尝试以符合逻辑的方式来标准化 DOM 事件 API。目前，主流的浏览器都实现了 DOM2 事件系统的核心部分。浏览器事件系统非常复杂，即使所有的主流浏览器都实现了 DOM2，规范也没有涵盖到所有的事件类型。DOM3 新增的事件 API 又让这些问题更加复杂了。

DOM2 将事件流分为事件捕获、到达目标（即实际的目标元素接收到事件）和事件冒泡（即事件向上层节点传递）3 个阶段。事件捕获为提前拦截事件提供了可能，最迟要在事件冒泡阶段响应事件。为响应事件而调用的事件处理函数称为事件处理程序（或事件监听器）。

React 元素的事件处理和 DOM 元素的事件处理很相似。但是，React 事件的命名方式采用小驼峰式。对于由多个英文单词组成的事件名称的第一个英文单词全部小写，而从第二个单词起的每个单词首字母大写。而且，使用 JSX 语法时需要传入一个函数（而不是字符串）作为事件处理函数。使用 React 事件时，不能通过返回 false 的方式阻止其默认行为，而必须显式地使用 preventDefault 来阻止。一般不需要使用 addEventListener 为已创建的 DOM 元素添加监听器，只需要在该元素初始渲染的时候添加监听器即可。定义类组件时，通常要将事件处理函数声明为类中的方法。

必须谨慎对待 JSX 回调函数中的 this。在 JavaScript 中，class 的方法默认不会绑定 this。例如，如果没有绑定 this.handleClick 并把它传入了 onClick，即 onClick={this.handleClick}；在调用这个函数的时候，this 的值为 undefined。在通常情况下，如果没有在方法后面添加 ()，就应该为这个方法绑定（bind）this。还可以使用实验性的类 field 语法或回调函数避免使用绑定。如果回调函数作为 props 传入子组件时，这些组件可能会进行额外的重新渲染。建议在构造方法中绑定或使用类 field 语法来避免这类性能问题。

在循环中，通常需要为事件处理函数传入额外的参数。例如，若 id 是要删除那一行的 ID，以下两种方式都可以向事件处理函数传入参数：

```
<button onClick={(e) => this.deleteRow(id, e)}>Delete Row</button>
<button onClick={this.deleteRow.bind(this, id)}>Delete Row</button>
```

上述两种方式是等价的，分别通过箭头函数（使用了=>）和 Function.prototype.bind() 方法（方法绑定）来实现。如果通过箭头函数的方式，事件对象必须显式地进行传递，React 的事件对象 e 会被作为第二个参数传递。通过方法绑定的方式，事件对象以及更多的参数将会被隐式地进行传递。

3.1.2 合成事件

合成事件即自定义事件。事件合成除了可以处理兼容性问题，还可以用来自定义高级事件，如 onChange 事件（表单元素统一的值变动事件）。虚拟 DOM 抽象了跨平台的渲染方式，SyntheticEvent 实例将被传递给事件处理函数，它是浏览器的原生事件的跨浏览器包装器。除兼容所有浏览器外，它还拥有和浏览器原生事件相同的接口，包括 stopPropagation() 方法和 preventDefault() 方法等。SyntheticEvent 是合并而来。这意味着 SyntheticEvent 对象可能会被重用，而且在事件回调函数被调用后，所有的属性都会无效。

大部分事件最终绑定到了 Document，而不是 DOM 节点本身。这就简化了 DOM 事件处理逻辑，减少了内存开销。不同类型的事件有不同的优先级，如高优先级的事件可以中断渲染，让用户代码可以及时响应用户交互。

3.1.3 支持的事件类型

常见的鼠标事件有 onClick、onContextMenu、onDoubleClick、onDrag、onDragEnd、onDragEnter、onDragExit、onDragLeave、onDragOver、onDragStart、onDrop、onMouseDown、onMouseEnter、onMouseLeave、onMouseMove、onMouseOut、onMouseOver 和 onMouseUp 等。其中 onMouseEnter 和 onMouseLeave 事件从离开的元素向进入的元素传播，不是正常的冒泡，也没有捕获阶段。

常见的焦点事件 onFocus、onBlur 在 React DOM 上的所有元素都有效，不只是表单元素。onFocus 事件在元素（或其内部某些元素）聚焦时被调用。onBlur 事件在元素（或其内部某些元素）失去焦点时被调用。可以使用 currentTarget 和 relatedTarget 来区分聚焦和失去焦点是否来自父元素外部。

常见的键盘事件有 onKeyDown、onKeyPress 和 onKeyUp。常见的图像事件有 onLoad 和 onError。

常见的指针事件（并非每个浏览器都支持指针事件）有 onPointerDown、onPointerMove、onPointerUp、onPointerCancel、onGotPointerCapture、onLostPointerCapture、onPointerEnter、onPointerLeave、onPointerOver 和 onPointerOut。其中 onPointerEnter 和 onPointerLeave 事件从离开的元素向进入的元素传播，不是正常的冒泡，也没有捕获阶段。

常见的表单事件有 onChange、onInput、onInvalid、onReset 和 onSubmit 等。常见的选择事件有 onSelect。常见的剪贴板事件有 onCopy、onCut 和 onPaste。常见的复合（Composition）事件有 onCompositionEnd、onCompositionStart 和 onCompositionUpdate。常见的通用事件有 onError 和 onLoad。常见的触摸事件有 onTouchCancel、onTouchEnd、onTouchMove 和 onTouchStart。

常见的 UI 事件有 onScroll。注意，从 React 17.0 开始，onScroll 事件在 React 中不再冒泡。这与浏览器的行为一致，并且避免了当一个嵌套且可滚动的元素在其父元素触发事件时造成的混乱。常见的滚轮事件有 onWheel。

常见的媒体事件有 onAbort、onCanPlay、onCanPlayThrough、onDurationChange、onEmptied、onEncrypted、onEnded、onError、onLoadedData、onLoadedMetadata、onLoadStart、onPause、onPlay、onPlaying、onProgress、onRateChange、onSeeked、onSeeking、onStalled、onSuspend、onTimeUpdate、onVolumeChange 和 onWaiting。

常见的动画事件有 onAnimationStart、onAnimationEnd 和 onAnimationIteration。常见的过渡事件有 onTransitionEnd。还有一些其他事件，如 onToggle。

3.2 鼠标事件处理

3.2.1 开发示例

在项目 firstreact 根目录下创建 eventexample 子目录，在 firstreact\eventexample 目录下创建文件 eventexample.html，代码与例 1-3 所示的代码相同。在 firstreact/eventexample 目录下创建文件 index.js，代码如例 3-1 所示。

【例 3-1】 在 firstreact\eventexample 目录下创建的文件 index.js 的代码。

```
const divReact = document.getElementById('root');
const infoMap= {
    clickAlertInfo:'发生了鼠标单击事件',
    linkPreventInfo:'即将访问 React 官网',
    reactWebsiteInfo:'https://reactjs.org/',
    gotoReactWebsiteInfo:'访问 React 官网',
    button2AlertInfo:'类事件处理正常',
    button2Value:'类的事件',
    callbackAlertInfo:'使用了箭头函数',
    callbackValue:'回调操作',
    eventWithPropsAlertInfo:'传递的参数为:name=',
```

```
            eventWithPropsValue:'带参数的事件处理',
            onOffStateAlertInfo:'现有的状态为',
            onOffButtonValue1:'开关按钮（初始为开）',
            onOffButtonValue2:'开关按钮（初始为关）',
            titleInfo:'事件处理的简单示例',
            button1Info:'鼠标单击事件示例：',
            button1Value:'单击鼠标事件',
            linkExInfo:'超链接访问示例：',
            button2Info:'类的事件处理示例：',
            callbackInfo:'回调操作示例：',
            eventWithPropsInfo:'带参数的事件处理示例：',
            onOffStateInfo:'bool 状态的变化示例：',
            preventDefaultInfo:'使用 preventDefault 阻止默认行为',
            actionLinkInfo:'单击 Link'
        }
        {/*鼠标单击事件处理*/}
        function onBtnClick() {
            alert(infoMap.clickAlertInfo);
        }
        {/*阻止事件的默认行为*/}
        function PreventLink() {
            function handleClick(e) {
                alert(infoMap.linkPreventInfo);
            }
            return (
                <a href={infoMap.reactWebsiteInfo} onClick={handleClick}>
                    {infoMap.gotoReactWebsiteInfo}
                </a>
            );
        }
        class BtnClickComp extends React.Component {
            constructor(props) {
                super(props);
            }
            handleClick2() {
                alert(infoMap.button2AlertInfo);
            }
            render() {
                return (
                    <button onClick={this.handleClick2}>
                        {infoMap.button2Value}
                    </button>
                );
            }
        }
        {/*使用了箭头函数，回调*/}
        class CallBackBtnClickComp extends React.Component {
            constructor(props) {
                super(props);
```

```jsx
        }
        handleClick = () => {
            alert(infoMap.callbackAlertInfo);
        };
        render() {
            // 此语法确保handleClick()方法内的this已被绑定
            return (
                <button onClick={(e) => this.handleClick(e)}>
                    {infoMap.callbackValue}
                </button>
            );
        }
}
{/*在事件处理方法中传递参数*/}
class ParamsEventComp extends React.Component {
    constructor(props) {
        super(props);
        this.name = props.name ;
    }
    passParamsClick(name, e) {//事件对象e要放在最后
        e.preventDefault();
        alert(infoMap.eventWithPropsAlertInfo+" "+name);
    }
    render() {
        return (
            <div>
                {/* 通过箭头函数方法传递参数 */}
                <button onClick={(e) => this.passParamsClick(this.name, e)}>
                    {infoMap.eventWithPropsValue}
                </button>
            </div>
        );
    }
}
class ToggleBtnComp extends React.Component {
    constructor(props) {
        super(props);
        this.isToggleOn=props.isToggleOn;
        //为了在回调中使用this,这里的绑定是必不可少的
        this.handleClick = this.toggleBtnClick.bind(this);
    }
    toggleBtnClick(isToggleOn, e) {
        e.preventDefault();
        alert(infoMap.onOffStateAlertInfo+" " + this.isToggleOn );
        this.isToggleOn = ! this.isToggleOn;
    }
    render() {
        return (
            <button onClick={(e)=>this.toggleBtnClick(this.isToggleOn, e)}>
```

```
                {this.isToggleOn ? infoMap.onOffButtonValue1 : infoMap.
onOffButtonValue2}
            </button>
        );
    }
}
function ActionLink() {
    //e 是合成事件，React 根据 W3C 规范来定义合成事件，开发时无须考虑跨浏览器兼容问题
    function handleClick(e) {
        e.preventDefault();
        alert(infoMap.actionLinkInfo);
    }
    return (
        <a href="#" onClick={handleClick}>
            {infoMap.actionLinkInfo}
        </a>
    );
}
const exampleEvent = (
    <span>
        <h2 align="center">{infoMap.titleInfo}</h2>
        <span>{infoMap.button1Info}</span>
<button onClick={onBtnClick}>{infoMap.button1Value}</button>
        <hr/>
        <span>{infoMap.linkExInfo}</span><PreventLink/>
        <hr/>
        <span>{infoMap.button2Info}</span><BtnClickComp/>
        <hr/>
        <span>{infoMap.callbackInfo}</span><CallBackBtnClickComp/>
        <hr/>
        <span>{infoMap.eventWithPropsInfo}</span><ParamsEventComp name=
{'zd'}/>
        <hr/>
        <span>{infoMap.onOffStateInfo}</span><ToggleBtnComp isToggleOn=
{true}/>
        <hr/>
        <span>{infoMap.preventDefaultInfo}</span><ActionLink/>
    </span>
);
ReactDOM.createRoot(divReact).render(exampleEvent);
```

3.2.2 运行效果

运行文件 eventexample.html，效果如图 3-1 所示。

单击图 3-1 所示中的"单击鼠标事件"按钮，弹出如图 3-2 所示的对话框。单击图 3-1 所示中的超链接"访问 React 官网"，弹出如图 3-3 所示的对话框。单击该对话框中的"确定"按钮，结果如图 3-4 所示。单击图 3-1 所示中的"类的事件"按钮，弹出如图 3-5 所示

的对话框。单击图 3-1 所示中的"回调操作"按钮，弹出如图 3-6 所示的对话框。单击图 3-1 所示中的"带参数的事件处理"按钮，弹出如图 3-7 所示的对话框。单击图 3-1 所示中的"开关按钮（初始为开）"按钮，弹出如图 3-8 所示的对话框。单击图 3-1 所示中的超链接"单击 Link"，弹出如图 3-9 所示的对话框。

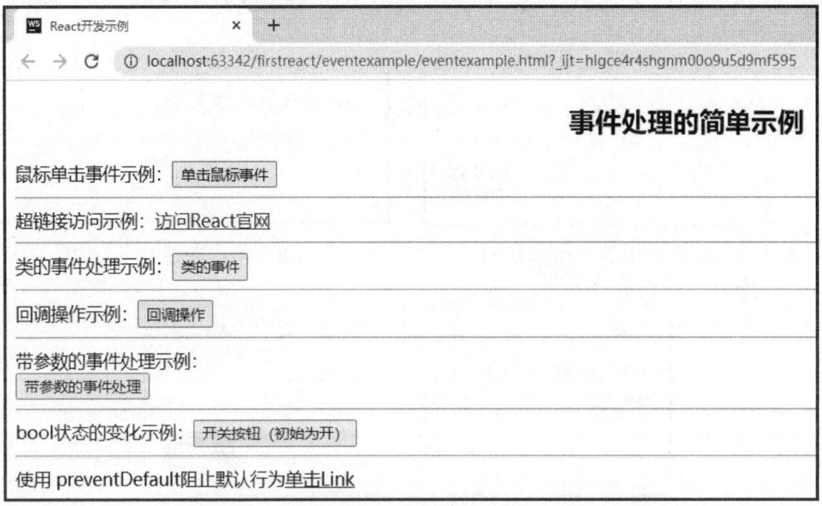

图 3-1　运行文件 eventexample.html 的效果

图 3-2　单击图 3-1 所示中的"单击
　　　　鼠标事件"按钮的效果

图 3-3　单击图 3-1 所示中的超链接
　　　　"访问 React 官网"的效果

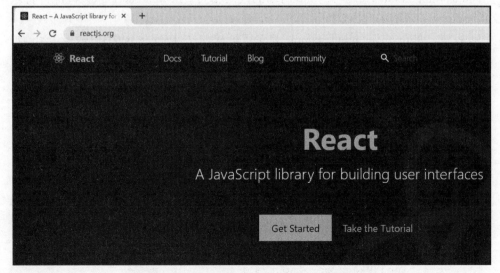

图 3-4　单击图 3-3 所示中的"确定"按钮的效果

图 3-5　单击图 3-1 所示中的"类的事件"按钮的效果　　图 3-6　单击图 3-1 所示中的"回调操作"按钮的效果

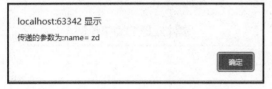

图 3-7　单击图 3-1 所示中的"带参数的　　　　图 3-8　单击图 3-1 所示中的"开关按钮
　　　　事件处理"按钮的效果　　　　　　　　　　　　　（初始为开）"按钮的效果

图 3-9　单击图 3-1 所示中的超链接"单击 Link"的效果

3.3　焦点事件处理

3.3.1　开发示例

在 firstreact 根目录下创建 inputexample 子目录，在 firstreact\inputexample 目录下创建文件 inputexample.html，代码与例 1-3 所示的代码相同。在 firstreact\inputexample 目录下创建文件 index.js，代码如例 3-2 所示。

【例 3-2】　在 firstreact\inputexample 目录下创建的文件 index.js 的代码。

```
const divReact = document.getElementById('root');
const infoMap= {
    input1focusAlertInfo:'第一个文本框获得焦点',
    input1changeAlertInfo:'第一个文本框内容变为',
    input1blurAlertInfo:'第一个文本框失去焦点',
    input2focusAlertInfo:'第二个文本框获得焦点',
    input2NotFocusAlertInfo:'第二个文本框失去焦点',
    titleInfo:'文本框事件简单示例',
}
class InputListenerComp extends React.Component {
    constructor(props) {
        super(props);
        this.inputVal=props.inputVal;
        this.inputTextFocus = this.inputTextFocus.bind(this);
        this.inputTextChange = this.inputTextChange.bind(this);
```

```
            this.inputTextBlur = this.inputTextBlur.bind(this);
        }
        inputTextFocus(inputVal) {
            alert(infoMap.input1focusAlertInfo);
        }
        inputTextChange(inputVal) {
            alert(infoMap.input1changeAlertInfo+inputVal);
        }
        inputTextBlur(inputVal) {
            alert(infoMap.input1blurAlertInfo);
        }
        render() {
            const inputVal = this.inputVal;
            return (
                <input type="text" value={inputVal}
                    onFocus={this.inputTextFocus(inputVal)}
                    onChange={this.inputTextChange(inputVal)}
                    onBlur={this.inputTextBlur(inputVal)} />
            );
        }
    }
    class InputFocusComp extends React.Component {
        constructor(props) {
            super(props);
            this.inputTextFocus = this.inputTextFocus.bind(this);
        }
        inputTextFocus(inputTextFocus) {
            if (inputTextFocus) {
                alert(infoMap.input2focusAlertInfo);
                inputTextFocus =! inputTextFocus;
            }
            alert(infoMap.input2NotFocusAlertInfo);
        }
        render() {
            return (
                <input type="text" onFocus={this.inputTextFocus} />
            );
        }
    }
    const exampleInputEvent = (
        <span>
            <h2 align="center">{infoMap.titleInfo}</h2>
            {/*注意"true"和 true 的差别*/}
            <InputListenerComp inputVal={"true"}/>
            <InputFocusComp  inputTextFocus={true}/>
        </span>
    );
    ReactDOM.createRoot(divReact).render(exampleInputEvent);
```

3.3.2 运行效果

运行文件 inputexample.html，效果如图 3-10 所示。单击图 3-10 所示中的"确定"按钮，弹出如图 3-11 所示的对话框。单击图 3-11 所示中的"确定"按钮，弹出如图 3-12 所示的对话框。单击图 3-12 所示中的"确定"按钮，效果如图 3-13 所示。单击图 3-13 所示中的第二个文本框（横向从左至右），弹出如图 3-14 所示的对话框。单击图 3-14 所示中的"确定"按钮，弹出如图 3-15 所示的对话框。单击图 3-15 所示中的"确定"按钮，弹出如图 3-16 所示的对话框。之后在图 3-15 和图 3-16 之间来回转换。

图 3-10　运行文件 inputexample.html 的效果

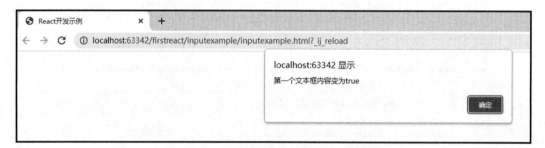

图 3-11　单击图 3-10 所示中的"确定"按钮的效果

图 3-12　单击图 3-11 所示中的"确定"按钮的效果

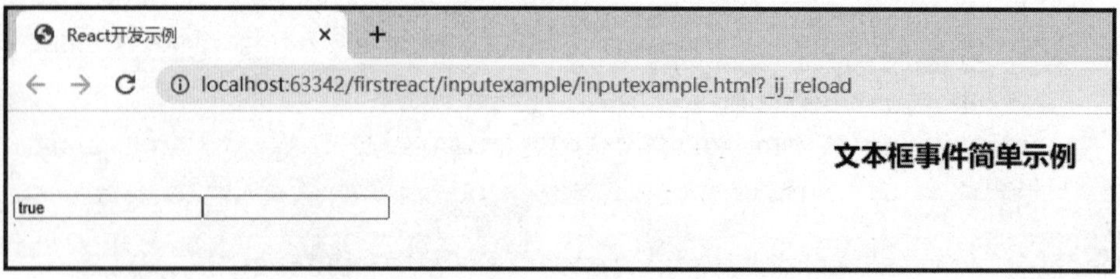

图 3-13　单击图 3-12 所示中的"确定"按钮的效果

图 3-14　单击图 3-13 所示中的第二个文本框（横向从左至右）的效果

图 3-15　单击图 3-14 所示中的"确定"按钮的效果

视频讲解

图 3-16　单击图 3-15 所示中的"确定"按钮的效果

3.4　键盘事件处理

3.4.1　开发示例

在 firstreact 根目录下创建 keyboardeventexample 子目录，在 firstreact\keyboardeventexample 目录下创建文件 keyboardeventexample.html，代码与例 1-3 所示的代码相同。在 firstreact\keyboardeventexample 目录下创建文件 index.js，代码如例 3-3 所示。

【例 3-3】　在 firstreact\keyboardeventexample 目录下创建的文件 index.js 的代码。

```
const divReact = document.getElementById('root');
const infoMap={
    msgInit:'通过键盘在文本框中输入字符后在浏览器 Console 中显示对应的 ASCII 码值',
    alertInfo:'按回车键，msg 值为：'
}
class KeyBind extends React.Component {
    constructor(props) {
        super(props)
        this.state = {
```

```
            msg: infoMap.msgInit
        }
    }
    keyUp = (e) => {
        console.log(e.keyCode)
        if (e.keyCode === 13) {
            alert(infoMap.alertInfo+e.target.value)
        }
    }
    render() {
        return (
            <div>
                <h2>{infoMap.msgInit}</h2>
                <input onKeyUp={this.keyUp}/>
            </div>
        )
    }
}
const keyboardEx = (
    <span>
        <KeyBind/>
    </span>
)
ReactDOM.createRoot(divReact).render(keyboardEx);
```

3.4.2 运行效果

运行文件 keyboardeventexample.html，效果如图 3-17 所示。通过在图 3-17 所示的文本框中输入字符（如"1""a"），在浏览器的 Console 中输出字符对应的 ASCII 码值（如 49、65），效果如图 3-18 所示。在图 3-18 所示中，通过按回车键，弹出的对话框（包含文本框中存有的内容，如"1a"）如图 3-19 所示。

图 3-17 运行文件 keyboardeventexample.html 的效果

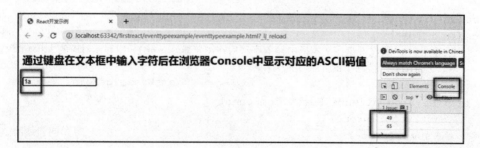

图 3-18 通过键盘在文本框中输入字符后在浏览器 Console 中显示对应的 ASCII 码值

图 3-19　在图 3-18 所示中通过按回车键后弹出的对话框

3.5　图像事件处理

3.5.1　开发示例

在 firstreact 根目录下创建 imageeventexample 子目录，在 firstreact\imageeventexample 目录下创建文件 imageeventexample.html，代码与例 1-3 所示的代码相同。在 firstreact\imageeventexample 目录下创建文件 index.js，代码如例 3-4 所示。还需要在 firstreact\imageeventexample 目录下创建 img 子目录，并且在 firstreact\imageeventexample\img 目录下准备图片文件 1.jpg、2.jpg、3.jpg 和 4.jpg。

【例 3-4】　在 firstreact\imageeventexample 目录下创建的文件 index.js 的代码。

```
const divReact = document.getElementById('root');
const infoMap = {
    imgPath: './img',
    imgEx:'.jpg',
    file1Name:'1',
    file2Name:'2',
    file3Name:'3',
    file4Name:'4',
    alertInfo: '装载图像'
}
//图像 onLoad 事件处理函数（方法）
function loadPic(){
    alert(infoMap.alertInfo)
}
class Img extends React.Component{
    constructor(props) {
        super(props);
        this.state={
            listImg:[
                './img/1.jpg',
                './img/2.jpg',
                './img/3.jpg',
                './img/4.jpg',
            ],
            index:0
        }
    }
//计时器执行
```

```jsx
    indexChange(){
        if(this.state.index == this.state.listImg.length-1){
            this.setState({
                index:0
            })
        }else{
            this.setState({
                index:this.state.index+1
            })
            console.log(this.state.index); //在浏览器 Console 中输出图片序号
        }
    }
    componentDidMount(){
        setInterval(()=>{
            this.indexChange();
        },2000)
    }
    render(){
        let {listImg,index} = this.state;
        let imgList=listImg.map((item,imgIndex)=>{
            return
    <img src={item} key={imgIndex} style={{'display':index == imgIndex ?'block' : 'none'}} className='img' onLoad={loadPic}/>
        })
        let liList=listImg.map((item2,imgIndex2)=>{
            return
            <li key={imgIndex2} style={{'listStyleType':index == imgIndex2 ? 'initial' :'circle'}}></li>
        })
        return (
            <div>
                {imgList}
                <div>
                    <ul >
                        {liList }
                    </ul>
                </div>
            </div>
        )
    }
}
const loadImageEx = (
    <span>
        <Img/>
    </span>
)
ReactDOM.createRoot(divReact).render(loadImageEx);
```

3.5.2 运行效果

运行文件 imageeventexample.html，效果如图 3-20 所示。单击图 3-20 所示中的"确定"按钮，即可开始装载图像（图片），如图 3-21 所示。在装载后，显示轮播信息（如图 3-21 所示中的 4 个圆点对应的图片）。装载之后，就在 4 幅图片之间循环播放，如图 3-22 所示。

图 3-20 运行文件 imageeventexample.html 的效果

图 3-21 装载图像后显示的轮播信息

图 3-22 装载后在 4 幅图片之间播放的过程

习题 3

一、简答题
1. 简述对事件的理解。
2. 简述对合成事件的理解。

二、实验题
1. 实现鼠标事件处理的应用开发。
2. 实现焦点事件处理的应用开发。
3. 实现键盘事件处理的应用开发。
4. 实现图像事件处理的应用开发。

第 4 章

React条件渲染、列表和key

本章简要介绍条件渲染、列表和 key，再介绍条件渲染、列表和 key 等的应用开发。

4.1 React 条件渲染、列表和 key 概述

4.1.1 条件渲染

在 React 中，可以创建不同的组件来封装各种行为。React 会根据应用的不同状态，渲染对应状态下的部分内容。React 的条件渲染使用 JavaScript 运算符或者 if 条件运算符去创建元素来表现当前的状态，并根据它们来更新 UI 界面。

借助于短路逻辑与可以很方便地进行元素的条件渲染。因为在 JavaScript 中（和其他高级语言一样），表达式 true && expression 总是会返回 expression，而表达式 false && expression 总是会返回 false。在表达式中，如果&&左侧条件是 true，&&右侧的元素是 expression，就会被渲染；如果&&左侧条件是 false，那么 React 会忽略表达式及其对应的语句块并跳过它。JavaScript 中的短路逻辑或（||）运算符也和高级语言的用法一样。

另一种条件渲染的方法是使用 JavaScript 中的三目运算符 condition ? true : false。注意，如果条件变得过于复杂，可以考虑分解（提取）组件。在极少数情况下，若希望隐藏组件，则可以让 render()方法直接返回 null，而不进行任何渲染。

4.1.2 列表

如例 4-1 所示，使用 map()方法可以让数组中的每一项数值都变成原来数值的两倍，得到一个新的列表 doubled 并打印出 [2, 4, 6, 8, 10]。

【例 4-1】 map()方法的应用代码。

```
const numbers = [1, 2, 3, 4, 5];
```

```
const doubled = numbers.map((number) => number * 2);
console.log(doubled);
```

在 React 中，把数组转化为元素列表的过程与此类似。可以通过使用{}在 JSX 内构建一个元素列表（集合）。

如例 4-2 所示，使用 map()方法遍历 numbers 数组，并将数组中的每个数值变成在<li>标签和</li>标签之间的元素从而得到新的数组，再将得到的新数组赋值给 listItems，再把整个 listItems 插入<ul>标签和</ul>标签中，然后渲染进 DOM 中。

【例 4-2】 用 map()方法渲染数组的应用代码。

```
const numbers = [1, 2, 3, 4, 5];
const listItems = numbers.map((number) =>
  <li>{number}</li>
);
ReactDOM.render(
  <ul>{listItems}</ul>,
  document.getElementById('root')
);
```

如例 4-2 所示，通常需要在一个组件中渲染元素列表（如数组）。还可以把例 4-2 重构成一个组件，这个组件接收 numbers 数组作为参数并输出一个元素列表。

4.1.3 key

通过数组构建一组子元素时，React 预期每个元素都有 key 属性。有了 key 属性，React 才能更高效地更新 DOM。由于 key 可以帮助 React 识别哪些元素发生了变化，因此应当给数组中的每一个元素赋予一个确定的标识。例如，在 map()方法中的元素需要设置 key 属性。

一个元素的 key 最好是这个元素在列表中拥有的一个独一无二的字符串。通常，可以使用数据中的 id 来作为元素的 key。

如果列表元素的顺序可能会变化，不建议使用索引作为 key 的值，因为这样做会导致性能变差，还可能引起组件状态的问题。如果不指定 key 值，React 会默认使用索引作为列表元素的 key 值。

同一数组元素之间的 key 值应该是独一无二的。然而并不需要 key 值是全局唯一的。两个不同的数组中，不同元素可以使用相同的 key 值，只要能区分同一数组内的兄弟元素即可。

视频讲解

4.2 条件渲染的应用开发

4.2.1 开发示例

在 firstreact 根目录下创建 ifexample 子目录，在 firstreact\ifexample 目录下创建文件 ifexample.html，代码与例 1-3 所示的代码相同。在 firstreact\ifexample 目录下创建文件

index.js，代码如例 4-3 所示。

【例 4-3】 在 firstreact\ifexample 目录下创建的文件 index.js 的代码。

```
const divReact = document.getElementById('root');
const infoMap= {
    loggedInInfo:'欢迎登录此系统。',
    loggedOutInfo:'请先登录系统。',
    loginButtonInfo:'登录',
    logoutButtonInfo:'注销',
    stateInfo:'现在的状态为',
    scoreInfo:'得分',
    passInfo:'分，通过考试。',
    notPassInfo:'分，没有通过考试。',
    passScore:60,
    btn2info:'按钮的使用 2：',
    cool:'今天真冷。',
    warm:'今天气温舒适。',
    hot:'今天有点热。',
    veryHot:'今天真热。',
    how:'今天气温如何？',
    coolStandard:0,
    warmStandard:25,
    hotStandard:38,
    title1info:'if 条件简单示例',
    trueInfo:'初始条件为 true 的结果：',
    falseInfo:'初始条件为 false 的结果：',
    title2info:'按钮的使用 1：',
    title3info:'与(&&)运算和或(||)运算的简单示例',
    title4info:'三目运算简单示例',
    temperature:25,
    temperatureUnit:'度，',
}
function UserLoggedInComp(props) {
    return <span>{infoMap.loggedInInfo}</span>;
}
function UserLoggedOutComp(props) {
    return <span>{infoMap.loggedOutInfo}</span>;
}
//用 if 语句进行条件渲染
function UserLoggedComp(props) {
    const isLogged = props.isLogged;
    if(isLogged) {
        return <UserLoggedInComp/>;
    } else {
        return <UserLoggedOutComp/>;
    }
}
//登录按钮组件
function LoginButton(props) {
    return (
```

```jsx
            <button onClick={props.onClick}>
                {infoMap.loginButtonInfo}
            </button>
        );
    }
    //注销按钮组件
    function LogoutButton(props) {
        return (
            <button onClick={props.onClick}>
                {infoMap.logoutButtonInfo}
            </button>
        );
    }
    //条件登录控件组件
    class LoginControl extends React.Component {
        constructor(props) {
            super(props);
            this.handleLoginClick = this.handleLoginClick.bind(this);
            this.handleLogoutClick = this.handleLogoutClick.bind(this);
            this.isLoggedIn=props.isLoggedIn;
        }
        handleLoginClick() {
           this.isLoggedIn= !this.isLoggedIn;
            alert(infoMap.stateInfo+this.isLoggedIn);
        }
        handleLogoutClick() {
            this.isLoggedIn=!this.isLoggedIn;
            alert(infoMap.stateInfo+this.isLoggedIn);
        }
        render() {
            let button;
            if (this.isLoggedIn) {
                button = <LogoutButton onClick={this.handleLogoutClick} />;
            } else {
                button = <LoginButton onClick={this.handleLoginClick} />;
            }
            return (
                <div>
                    <UserLoggedComp isLogged={this.isLoggedIn} />
                    {button}
                </div>
            );
        }
    }
    const score = [90, 60, 50];
    function PassTestComp(props) {
        return (
        <p>
            {infoMap.scoreInfo}<b>{props.score}</b>{infoMap.passInfo}
        </p>);
    }
```

```jsx
function NotPassTestComp(props) {
    return (
        <p>
            {infoMap.scoreInfo}<b>{props.score}</b>{infoMap.notPassInfo}
        </p>);
}
//使用&&和||
function TestPass(props) {
    const score = props.score;
    return (
        <div>
            {
                (score >= infoMap.passScore &&
                  <PassTestComp score={score}/>
                )
                 ||
                (score < infoMap.passScore &&
                    <NotPassTestComp score={score}/>
                )
            }
        </div>
    );
}
//运用三目运算
function TestPass2(props) {
    const score = props.score;
    return (
        <div>
            {
                (
                    score >= infoMap.passScore ?
                        <PassTestComp score={score}/>
                        :
                        <NotPassTestComp score={score}/>
                )
            }
        </div>
    );
}
function Banner(props) {
    if (!props.isBanner) {
        return null;
    }
    return (
        <div className="banner">
            {infoMap.btn2info}
        </div>
    );
}
class BannerComp extends React.Component {
    constructor(props) {
```

```
            super(props);
            this.showBanner=props.showBanner
            this.handleToggleClick = this.handleToggleClick.bind(this);
        }
        handleToggleClick() {
            alert(this.showBanner)
        }
        render() {
            return (
                <div>
                    <Banner isBanner={this.showBanner} />
                    <button onClick={this.handleToggleClick}>
                        {this.showBanner ? 'Hide' : 'Show'}
                    </button>
                </div>
            );
        }
    }
    function TempLevel(props) {
        if (props.wlevel <= infoMap.coolStandard) {
            return <p>{infoMap.cool}</p>;
        } else if((props.wlevel > infoMap.coolStandard) && (props.wlevel <= infoMap.warmStandard)) {
            return <p>{infoMap.warm}</p>;
        } else if((props.wlevel > infoMap.warmStandard) && (props.wlevel <= infoMap.hotStandard)) {
            return <p>{infoMap.hot}</p>;
        } else if (props.wlevel > infoMap.hotStandard)   {
            return <p>{infoMap.veryHot}</p>;
        } else {
            return <p>{infoMap.how}</p>;
        }
    }
    class WaterTempComp extends React.Component {
        constructor(props) {
            super(props);
            this.temperature=props.temperature;
        }
        render() {
            const temperature = this.temperature;
            return (
                <TempLevel wlevel={parseFloat(temperature)}/>
            );
        }
    }
    const exampleCondition = (
        <span>
            <h3>{infoMap.title1info}</h3>
            {infoMap.trueInfo}<UserLoggedComp isLogged ={true}/>
            <br/>
            {infoMap.falseInfo}<UserLoggedComp isLogged ={false}/>
```

```
            <hr/>
            {infoMap.title2info}<LoginControl isLoggedIn={true}/>
            <LoginControl isLoggedIn={false}/>
            <hr/>
            <h3>{infoMap.title3info}</h3>
                <TestPass score={score[0]} />
                <TestPass score={score[1]} />
                <TestPass score={score[2]} />
            <hr/>
            <h3>{infoMap.title4info}</h3>
                <TestPass score={score[0]} />
                <TestPass score={score[1]} />
                <TestPass score={score[2]} />
            <hr/>
            <BannerComp showBanner={true}/>
            <hr/>
            {infoMap.temperature}{infoMap.temperatureUnit}<WaterTempComp
temperature={infoMap.temperature}/>
        </span>
    );
    ReactDOM.createRoot(divReact).render(exampleCondition);
```

4.2.2 运行效果

运行文件 ifexample.html，效果如图 4-1 所示。单击图 4-1 所示中的"注销"按钮，弹

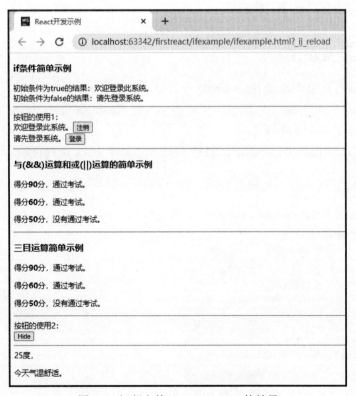

图 4-1 运行文件 ifexample.html 的效果

出如图 4-2 所示的对话框。单击图 4-1 所示中的"登录"按钮,弹出如图 4-3 所示的对话框。单击图 4-1 所示中的 Hide 按钮,弹出如图 4-4 所示的对话框。

图 4-2　单击图 4-1 所示中的"注销"按钮弹出的对话框

图 4-3　单击图 4-1 所示中的"登录"按钮弹出的对话框

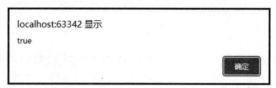

图 4-4　单击图 4-1 所示中的 Hide 按钮弹出的对话框

视频讲解

4.3　列表的应用开发

4.3.1　开发示例

在 firstreact 根目录下创建 listexample 子目录,在 firstreact\listexample 目录下创建文件 listexample.html,代码与例 1-3 所示的代码相同。在 firstreact\listexample 目录下创建文件 index.js,代码如例 4-4 所示。

【例 4-4】 在 firstreact\listexample 目录下创建的文件 index.js 的代码。

```
const divReact = document.getElementById('root');
const infoMap= {
    bookName:'书名————《 ',
    bookNameEnd:'》',
    author:'作者————',
    publisher:'出版社————',
    keywords:'关键词：',
    titleInfo:'List 简单示例',
}
const alpha = ['a', 'b'];
const alphaList = alpha.map(
    (alpha) => <li>{alpha}</li>
);
```

```
function MapList(props) {
    const alpha = props.alpha;
    const listAlpha = alpha.map(
        (alpha) => <li>{alpha.toUpperCase()} - {alpha}</li>
    );
    return (
        <ul>
            {listAlpha}
        </ul>
    );
}
const book1 ={
    name: 'Spring Boot 开发实战-微课视频版',
    author: '吴胜',
    publisher:'清华大学出版社',
    skills: ['Spring Boot']
}
const book2={
    name: 'Spring Cloud 微服务开发实战-微课视频版',
    author: '吴胜',
    publisher:'清华大学出版社',
    skills: ['Spring Cloud','微服务']
}
const book3={
    name: 'Spring Boot 区块链应用开发入门-微课视频版',
    author: '吴胜',
    publisher:'清华大学出版社',
    skills: ['区块链', 'Spring Boot']
}
const book4={
    name: '微信小程序云开发——Spring Boot+Node.js 项目实战'
    author: '吴胜',
    publisher:'清华大学出版社',
    skills: [ '微信小程序云开发','Spring Boot']
}
const book5={
    name: '微信小程序开发基础',
    author: '吴胜',
    publisher:'清华大学出版社',
    skills: ['微信小程序']
}
function BookList(props) {
    const books = props.books;
    const bookName = books.name;
    const author = books.author;
    const publisher = books.publisher;
    const skills = books.skills;
```

```
        const bookSkills = skills.map(
            (skills) => <li>{skills}</li>
        );
        return (
            <ul>
                <li>{infoMap.bookName}{bookName}{infoMap.bookNameEnd}</li>
                <li>{infoMap.author}{author}</li>
                <li>{infoMap.publisher}{publisher}</li>
                <li>{infoMap.keywords}
                    <ul>
                        {bookSkills}
                    </ul>
                </li>
            </ul>
        );
    }
    function BookComp(props) {
        const booklist = props.bookList;
        const booklistMap = booklist.map(
            (bookList) => <BookList books={bookList}></BookList>
        );
        return (
            <ul>
                {booklistMap}
            </ul>
        );
    }
    const bookList = [
        book1,book2,book3,book4,book5
    ];
    const exampleList = (
        <span>
            <h3 align="center">{infoMap.titleInfo}</h3>
            <ul>{alphaList}</ul>
            <hr/>
            <MapList alpha={alpha} />
            <hr/>
            <BookComp bookList={bookList} />
        </span>
    );
    ReactDOM.createRoot(divReact).render(exampleList);
```

4.3.2 运行效果

运行文件 listexample.html,效果如图 4-5 所示。

第4章 React条件渲染、列表和key

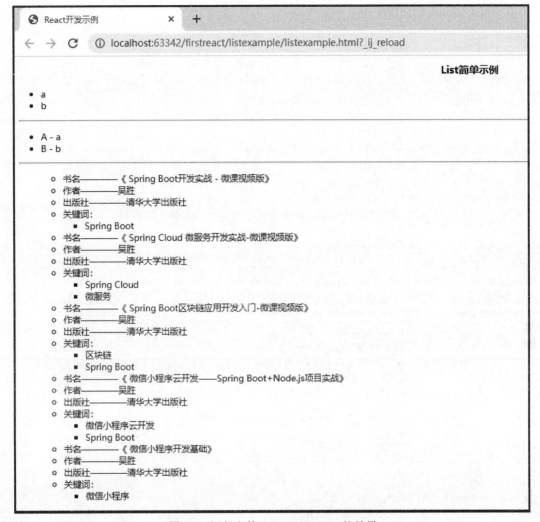

图 4-5 运行文件 listexample.html 的效果

4.4 key 的应用开发

4.4.1 开发示例

在 firstreact 根目录下创建 keyexample 子目录，在 firstreact\keyexample 目录下创建文件 keyexample.html，代码与例 1-3 所示的代码相同。在 firstreact\keyexample 目录下创建文件 index.js，代码如例 4-5 所示。

【例 4-5】 在 firstreact\keyexample 目录下创建的文件 index.js 的代码。

```
const divReact = document.getElementById('root');
const infoMap= {
    bookName:'书名————《 ',
    bookNameEnd:'》',
    author:'作者————',
```

```
    publisher:'出版社————',
    keywords:'关键词：',
    titleInfo:'key简单示例',
}
const alpha = ['a', 'b'];
//key值为i
var i=0;
const alphaList = alpha.map(
    (alpha) => <li key={i++}>{alpha}</li>
);
const id = new Array(3);
for(let i=0; i<3; i++) {
    id[i] = Math.round((Math.random()*Math.pow(10,10)));
}
const alphaList5 = alpha.map(
    (alpha, index) => <li key={id[index].toString()}>{index}——{alpha}</li>
);
i=0;
function MapList(props) {
    const alpha = props.alpha;
    const listAlpha = alpha.map(
        (alpha) => <li key={i++}>{alpha.toUpperCase()} - {alpha}</li>
    );
    return (
        <ul>
            {listAlpha}
        </ul>
    );
}
function BookList(props) {
    const books = props.books;
    const bookName = books.name;
    const author = books.author;
    const publisher = books.publisher;
    i=0
    const skills = books.skills;
    const bookSkills = skills.map(
        (skills) => <li key={i++}>{skills}</li>
    );
let j=0; //key值为j
    return (
        <ul>
            <li>{infoMap.bookName}{bookName}{infoMap.bookNameEnd}</li>
            <li>{infoMap.author}{author}</li>
            <li>{infoMap.publisher}{publisher}</li>
            <li key={j++}>{infoMap.keywords}
                <ul>
                    {bookSkills}
                </ul>
```

```
            </li>
        </ul>
    );
}
i=0;
function BookComp(props) {
    const booklist = props.bookList;
    const booklistMap = booklist.map(
        (bookList) => <BookList books={bookList} key={i++}></BookList>
    );
    return (
        <ul>
            {booklistMap}
        </ul>
    );
}
const book1 ={
    name: 'Spring Boot 开发实战-微课视频版',
    author: '吴胜',
    publisher:'清华大学出版社',
    skills: ['Spring Boot']
}
const book2={
    name: 'Spring Cloud 微服务开发实战-微课视频版',
    author: '吴胜',
    publisher:'清华大学出版社',
    skills: ['Spring Cloud','微服务']
}
const book3={
    name: 'Spring Boot 区块链应用开发入门-微课视频版',
    author: '吴胜',
    publisher:'清华大学出版社',
    skills: ['区块链', 'Spring Boot']
}
const bookList = [
    book1,book2,book3
];
const numbers = [1, 2, 3];
const listItems = numbers.map((number) =>
    <li key={number.toString()}>
        {number}
    </li>
);
const exampleKey = (
    <span>
        <h2 align="center">{infoMap.titleInfo}</h2>
        <ul>{alphaList}</ul>
        <hr/>
        <ul>{alphaList5}</ul>
```

```
        <hr/>
        <MapList alpha={alpha} />
        <hr/>
        <ul>{listItems}</ul>
        <hr/>
        <BookComp bookList={bookList} />
    </span>
);
ReactDOM.createRoot(divReact).render(exampleKey);
```

4.4.2 运行效果

运行文件 keyexample.html，效果如图 4-6 所示。

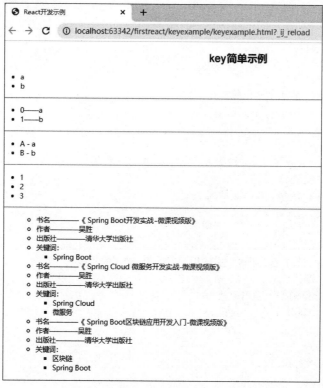

图 4-6　运行文件 keyexample.html 的效果

4.5　列表和 key 的综合应用

4.5.1 开发示例

在 firstreact 根目录下创建 exlistandkey 子目录，在 firstreact\exlistandkey 目录下创建文件 exlistandkey.html，代码与例 1-3 所示的代码相同。在 firstreact\exlistandkey 目录下创建文件 index.js，代码如例 4-6 所示。

【例 4-6】 在 firstreact\exlistandkey 目录下创建的文件 index.js 的代码。

```js
const divReact = document.getElementById('root');
const infoMap= {
    titleInfo:'列表和 key 简单示例',
    title1:'React.js 开发示例 1',
    content1:'React.js 组件的应用开发',
    title2:'React.js 安装方法 1',
    content2:'可以使用 npm 安装',
}
const numbers = [1, 2, 3];
function NumberList(props) {
    const numbers = props.numbers;
    const listItems = numbers.map((number) =>
        <li key={number.toString()}>{number}</li>
    );
    return (
        <ul>{listItems}</ul>
    )
}
function ListItem(props) {
    //不需要指定 key 的示例
    return <li>{props.value}</li>;
}
function NumberList2(props) {
    const numbers = props.numbers;
    const listItems = numbers.map((number) =>
        //key 应该在数组的上下文中被指定的示例
        <ListItem key={number.toString()} value={number} />
    );
    return (
        <ul>
            {listItems}
        </ul>
    );
}
function Blog(props) {
    const sidebar = (
        <ul>
            {props.posts.map((post) =>
                <li key={post.id}>
                    {post.title}
                </li>
            )}
        </ul>
    );
    const content = props.posts.map((post) =>
        <div key={post.id}>
            <h3>{post.title}</h3>
```

```
                <p>{post.content}</p>
            </div>
        );
    return (
        <div>
            {sidebar}
            <hr />
            {content}
        </div>
    );
}
const posts = [
    {id: 1, title: infoMap.title1, content: infoMap.content1},
    {id: 2, title: infoMap.title2, content: infoMap.content2}
];
const exampleList = (
    <span>
        <h3 align="center">{infoMap.titleInfo}</h3>
        <NumberList numbers={numbers} />
        <hr/>
        <NumberList2 numbers={numbers} />
        <hr/>
        <Blog posts={posts} />
    </span>
);
ReactDOM.createRoot(divReact).render(exampleList);
```

4.5.2 运行效果

运行文件 exlistandkey.html,效果如图 4-7 所示。

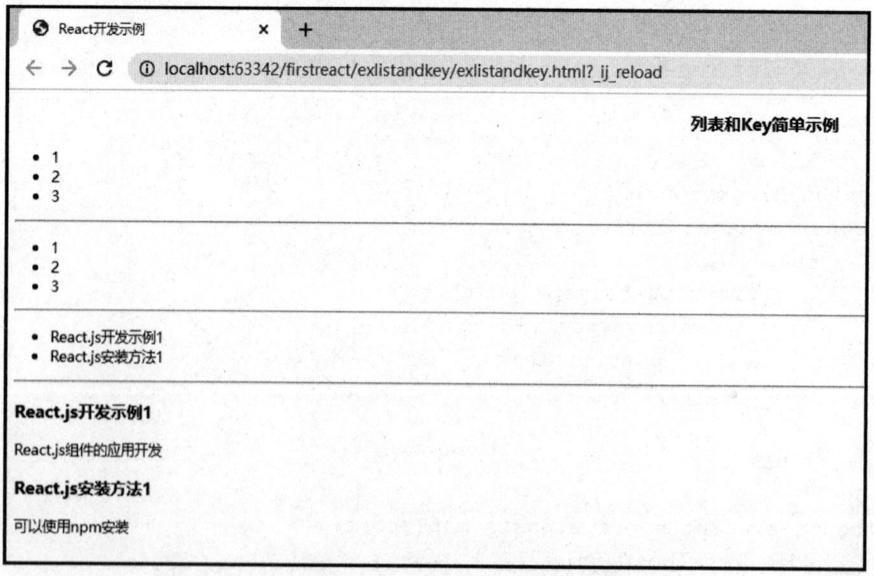

图 4-7 运行文件 exlistandkey.html 的效果

习题 4

一、简答题

1. 简述对条件渲染的理解。
2. 简述对列表的理解。
3. 简述对 key 的理解。

二、实验题

1. 实现条件渲染的应用开发。
2. 实现列表的应用开发。
3. 实现 key 的应用开发。

第 5 章

React状态管理

本章先简要介绍 state（状态）、setState()方法、forceUpdate()方法、状态提升等内容，再介绍状态的基础应用开发以及状态提升的应用开发等内容。

5.1 React 状态管理概述

5.1.1 state

组件中的 state 包含了随时可能发生变化的数据。state 与 props 类似，但 state 由用户自定义，是私有的，并且完全受控于当前组件。它是一个普通 JavaScript 对象。如果某些数据（值）未用于渲染或数据流，就不必将其设置为 state。此类数据可以在组件实例上定义。

如例 5-1 所示的类组件 Clock，在每次组件更新时，render()方法都会被调用，但只要在相同的 DOM 节点中渲染 <Clock />，就只有一个 Clock 组件实例被创建。

【例 5-1】类组件 Clock 的代码。

```
class Clock extends React.Component {
  render() {
    return (
      <div>
        <h1>Hello, world!</h1>
        <h2>It is {this.props.date.toLocaleTimeString()}.</h2>
      </div>
    );
  }
}
```

可以通过三个步骤向类组件中添加局部 state：把 render()方法中的 this.props 替换成

this.state；添加一个构造方法，然后在该方法中为 this.state 赋初值；转移属性（从组件被调用时的 props 转换到组件内）。将例 5-1 中的 date 从 props 移动到 state 中，经过转换后的带 state 的类组件 Clock 代码如例 5-2 所示。

【例 5-2】 转换后的带 state 的类组件 Clock 的代码。

```
class Clock extends React.Component {
  constructor(props) {
    super(props);
    this.state = {date: new Date()};
  }
  render() {
    return (
      <div>
        <h1>Hello, world!</h1>
        <h2>It is {this.state.date.toLocaleTimeString()}.</h2>
      </div>
    );
  }
}
```

5.1.2　setState()方法

setState(updater, [callback])方法简称为 setState()方法。将对组件 state 的更改排入队列，并通知 React 使用更新后的 state 重新渲染此组件及其子组件。这是用于更新 UI 界面以响应事件和处理服务器数据的主要方式。

可以在组件的构造方法中给 this.state 赋值。除了在组件构造方法中直接给 this.state 赋值之外，应使用 setState()方法来修改 state，而不要直接修改 state（如赋值语句）。同时，可以把多个 state 的 setState()方法调用合并成一个方法。

在调用 setState()方法时，React 会先把提供的对象合并到当前的 state，再分别调用 setState() 方法来单独地更新它们。

不管是父组件或是子组件，都无法知道某个组件是否有 state，并且它们也并不关心这个组件是函数组件还是类组件。组件可以选择把它的 state 作为 props 向下传递到子组件中，但是组件本身无法知道它是来自组件（或父组件）的 state 或 props，还是手动输入的。这通常会被叫作自上而下的单向数据流。任何 state 总是属于特定的组件，而且从该 state 派生的任何数据或 UI 界面只能影响组件树中低于它们的组件（如它们的子组件）。如果把一个以组件树想象成一个 props 的数据瀑布的话，那么每一个组件的 state 就像是在任意一点上给瀑布增加额外的水源，它只能向下流动。

在调用 setState()方法后，若立即读取 this.state，可能会有隐患。为了消除隐患，可以使用 componentDidUpdate()方法或者 setState()方法，这两种方式都可以保证在应用更新后触发。

除非 shouldComponentUpdate()方法返回 false，否则 setState()方法将始终执行重新渲染操作。如果使用可变对象，且无法在 shouldComponentUpdate() 方法中实现条件渲染，那么仅在新旧状态不同时调用 setState()方法，就可以避免不必要的重新渲染。

5.1.3 forceUpdate()方法

组件的 forceUpdate(callback)方法简称为 forceUpdate()方法。在默认情况下，当组件的 state 或 props 发生变化时，组件将重新渲染。如果 render()方法依赖于其他数据，就可以调用 forceUpdate() 方法强制让组件重新渲染。

调用 forceUpdate()方法将会导致组件调用 render()方法，此操作会跳过该组件的 shouldComponentUpdate() 方法。但其子组件会触发正常的生命周期方法，包括 shouldComponentUpdate() 方法。如果标签发生变化，React 仍将只更新 DOM。

通常，应避免使用 forceUpdate()方法，尽量在 render()方法中使用 this.props 和 this.state。

5.1.4 状态提升

当多个组件需要反映相同的变化数据时，可以将共享状态提升到最近的共同父组件中去。这和面向对象设计（如 Java 开发）时将子类共同内容抽取到父类中的思想类似。在 React 中，将多个组件中需要共享的 state 向上移动到它们最近的共同父组件中，便可实现多个组件共享 state。这就是所谓的"状态提升"。

React 中的任何可变数据应当只有一个相对应的唯一数据源。通常，state 都是先添加到需要渲染数据的组件中去。如果其他组件也需要这个 state，那么就可以将它提升至这些组件最近的共同父组件中。

虽然提升 state 方式比双向绑定方式需要编写更多的代码，但带来的好处是，排查和隔离 bug 所需的工作量将会变少。还可以使用自定义逻辑来拒绝或转换用户的输入。

在 UI 界面中发现错误时，可以检查问题组件的 props，并且按照组件树结构逐级向上搜寻，直到定位到负责更新 state 的组件。

视频讲解

5.2 状态的基础应用

5.2.1 开发示例

在 firstreact 根目录下创建 stateexample 子目录，在 firstreact\stateexample 目录下创建文件 stateexample.html，代码与例 1-3 所示的代码相同。在 firstreact\stateexample 目录下创建文件 index.js，代码如例 5-3 所示。

【例 5-3】 在 firstreact\stateexample 目录下创建的文件 index.js 的代码。

```
const divReact = document.getElementById('root');
const infoMap= {
    title1info:'状态和生命周期的应用',
    nowInfo:'现在时间是：',
    endInfo:'。',
    title2info:'数据流应用示例1',
    title3info: 'setState()方法示例',
    countInfo:'现在的计数是：',
```

```
        helloInfo:'Hello React!',
        btn1info:'箭头函数',
        btn2info:'bind()方法',
        title4info:'数据流应用示例2',
        title5info:'数据流应用示例3',
    }
    class ClockReactComp extends React.Component {
        constructor(props) {
            super(props);
            this.state = {date: new Date()};
        }
        //生命周期的方法
        componentDidMount() {
            this.timerId = setInterval(
                () => this.tick(),
                1000
            );
        }
        componentWillUnmount() {
            clearInterval(this.timerId);
        }
        tick() {
            this.setState({
                date: new Date()
            });
        }
        render() {
            return (
                <span>
                        <h4>{infoMap.title1info}</h4>
                      <p>{infoMap.nowInfo}{this.state.date.toLocaleTimeString()}
{infoMap.endInfo}</p>
                    </span>
                );
        }
    }
    function FormattedDate(props) {
        return <h4>{infoMap.nowInfo}{props.date.toLocaleTimeString()}{infoMap.
endInfo}</h4>;
    }
    class ClockReactComp2 extends React.Component {
        static defaultProps = {
            propsDate: new Date()
        };
        constructor(props) {
            super(props);
            this.state = {date: new Date()};
        }
        componentDidMount() {
```

```jsx
            this.timerId = setInterval(
                () => this.tick(),
                1000
            );
        }
        componentWillUnmount() {
            clearInterval(this.timerId);
        }
        tick() {
            this.setState({
                date: new Date()
            });
        }
        render() {
            return (
                <span>
                        <h3>{infoMap.title2info}</h3>
                        <FormattedDate date={this.state.date} />
                        <FormattedDate date={new Date()} />
                        <FormattedDate date={this.props.propsDate} />
                    </span>
            );
        }
    }
    class ClockReactComp3 extends React.Component {
        static defaultProps = {
            propsDate: new Date()
        };
        constructor(props) {
            super(props);
            this.state = {date: new Date()};
        }
        componentDidMount() {
            this.timerId = setInterval(
                () => this.tick(),
                1000
            );
        }
        componentWillUnmount() {
            clearInterval(this.timerId);
        }
        tick() {
            this.setState({
                date: new Date()
            });
        }
        render() {
            return (
                <span>
```

```jsx
                <FormattedDate date={this.state.date} />
            </span>
        );
    }
}
//数据流
function ClockDataFlow() {
    return (
        <div>
            <ClockReactComp3/>
            <ClockReactComp3/>
        </div>
    );
}
class ClockReactComp4 extends React.Component {
    static defaultProps = {
        propsDate: new Date()
    };
    constructor(props) {
        super(props);
        this.state = {date: new Date()};
    }
    componentDidMount() {
        this.timerId = setInterval(
            () => this.tick(),
            Math.ceil(Math.random()*9)*1000
        );
    }
    componentWillUnmount() {
        clearInterval(this.timerId);
    }
    tick() {
        this.setState({
            date: new Date()
        });
    }
    render() {
        return (
            <span>
                <FormattedDate date={this.state.date} />
            </span>
        );
    }
}
function ClockDataFlow2() {
    return (
        <div>
            <ClockReactComp4 />
            <ClockReactComp4 />
```

```jsx
            </div>
        );
    }
    class CountReactComp extends React.Component {
        constructor(props) {
            super(props);
            this.state = {
                count: 0
            };
        }
        componentDidMount() {
            this.timerId = setInterval(
                () => this.count(),
                1000
            );
        }
        count() {
            this.setState((prevState, props) => ({
                count: prevState.count + props.increment
            }));
        }
        componentWillUnmount() {
            clearInterval(this.timerId);
        }
        render() {
            return (
                <span>
                        <div>{infoMap.title3info}</div>
                        <div>{infoMap.countInfo}{this.state.count}{infoMap.endInfo}</div>
                    </span>
            );
        }
    }
    //注意代码的顺序
    //defaultProps 应用
    CountReactComp.defaultProps = {
        increment: 1
    };
    class ParamsEventComp extends React.Component {
        constructor(props) {
            super(props);
            this.state = {
                name: infoMap.helloInfo
            };
        }
        passParamsClick(name, e) {    //事件对象 e 要放在最后
            e.preventDefault();
            alert(name);
```

```jsx
        }
        render() {
            return (
                <div>
                    {/* 通过箭头函数方法传递参数 */}
                    <button onClick={(e)=>this.passParamsClick(this.state.name, e)}>
                        {infoMap.btn1info}
                    </button>
                </div>
            );
        }
    }
    class ParamsEventComp2 extends React.Component {
        constructor(props) {
            super(props);
            this.state = {
                name: infoMap.helloInfo
            };
        }
        passParamsClick(name, e) {
            e.preventDefault();
            alert(name);
        }
        render() {
            return (
                <div>
                    {/* 通过 bind()方法绑定传递参数 */}
                    <button onClick={this.passParamsClick.bind(this, this.state.name)}>
                        {infoMap.btn2info}
                    </button>
                </div>
            );
        }
    }
    const exampleState= (
        <span>
            <ClockReactComp/>
            <hr/>
            <ClockReactComp2/>
            <hr/>
            <h4>{infoMap.title4info}</h4>
            <ClockDataFlow />
            <hr/>
            <h4>数据流应用示例 3</h4>
            <ClockDataFlow2 />
            <hr/>
            <CountReactComp/>
```

```
            <hr/>
            <div>{infoMap.btn1info}</div>
            <ParamsEventComp />
            <hr/>
            <div>{infoMap.btn2info}</div>
            <ParamsEventComp2 />
        </span>
);
ReactDOM.createRoot(divReact).render(exampleState);
```

5.2.2 运行效果

运行文件 stateexample.html，效果如图 5-1 所示。注意，由于图 5-1 所示中输出的时间是运行程序的时间，读者在运行该程序时的结果会与图 5-1 不同，读者多次运行同一程序时的结果也不会相同，只要正确运行即可。单击图 5-1 所示中的"箭头函数"按钮，弹出的对话框如图 5-2 所示。单击图 5-1 中的"bind()方法"按钮，弹出的对话框如图 5-3 所示。

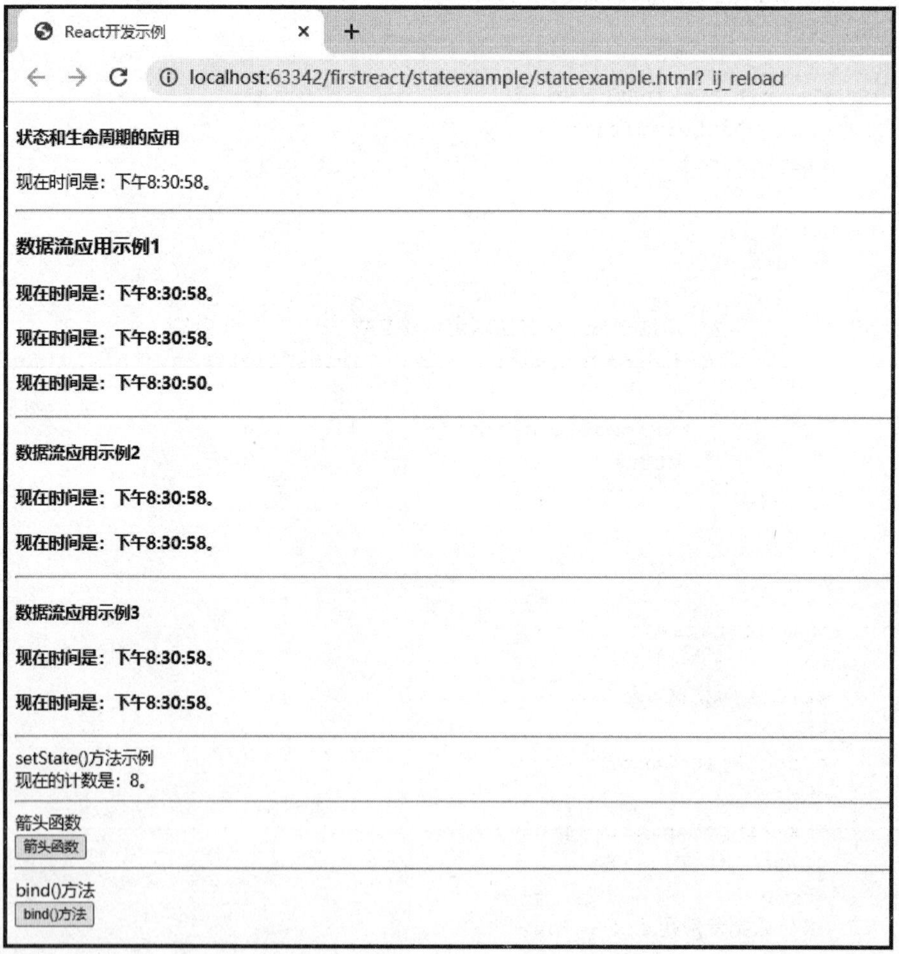

图 5-1 运行文件 stateexample.html 的效果

图 5-2　单击图 5-1 所示中的"箭头函数"按钮后弹出的对话框

图 5-3　单击图 5-1 所示中的"bind()方法"按钮后弹出的对话框

5.3　状态的提升应用

5.3.1　开发示例

在项目 firstreact 根目录下创建 advancedstateexample 子目录，在 firstreact\advancedstateexample 目录下创建文件 advancedstateexample.html，代码与如例 1-3 所示的代码相同。在 firstreact\advancedstateexample 目录下创建文件 index.js，代码如例 5-4 所示。

【例 5-4】　在 firstreact\advancedstateexample 目录下创建的文件 index.js 的代码。

```
const divReact = document.getElementById('root');
const infoMap= {
    cTempType:'摄氏度',
    fTempType:'华氏度',
    coolInfo:'今天真冷。',
    warmInfo:'今天气温舒适。',
    hotInfo:'今天有点热。',
    veryHotInfo:'今天真热。',
    howInfo:'今天气温如何？',
    cCoolStandard: 0,
    cWarmStandard: 20,
    cHotStandard: 38,
    fCoolStandard: 32,
    fWarmStandard: 68,
    fHotStandard: 100,
    outputInfo:'请输入气温（',
    outputInfoEnd:'）：',
    title1info:'状态提升示例',
    title2info:'示例1：两个文本框的气温数值相同（不考虑摄氏度和华氏度的区别）',
    title3info:'示例2：两个文本框的气温相等（考虑了摄氏度和华氏度之间的转换关系）',
}
const tempType = {
    c: infoMap.cTempType,
```

```
            f:infoMap.fTempType,
    };
    function TempLevel(props) {
        if(props.type === 'c') {
            return <TempCelsiusLevel wlevel={props.wlevel} />;
        } else if(props.type === 'f') {
            return <TempFahrenheitLevel wlevel={props.wlevel} />;
        } else {}
    }
    function TempCelsiusLevel(props) {
        if (props.wlevel <= infoMap.cCoolStandard) {
            return <p>{infoMap.coolInfo}</p>;
        } else if((props.wlevel > infoMap.cCoolStandard) && (props.wlevel<=
infoMap.cWarmStandard)) {
            return <p>{infoMap.warmInfo}</p>;
        } else if((props.wlevel > infoMap.cWarmStandard) && (props.wlevel<=
infoMap.cHotStandard)) {
            return <p>{infoMap.hotInfo}</p>;
        } else if(props.wlevel > infoMap.cHotStandard) {
            return <p>{infoMap.veryHotInfo}</p>;
        } else {
            return <p>{infoMap.howInfo}</p>;
        }
    }
    function TempFahrenheitLevel(props) {
        if (props.wlevel <= infoMap.fCoolStandard) {
            return <p>{infoMap.coolInfo}</p>;
        } else if((props.wlevel > infoMap.fCoolStandard) && (props.wlevel<=
infoMap.fWarmStandard)) {
            return <p>{infoMap.warmInfo}</p>;
        } else if((props.wlevel > infoMap.fWarmStandard) && (props.wlevel<=
infoMap.fHotStandard)) {
            return <p>{infoMap.hotInfo}</p>;
        } else if(props.wlevel > infoMap.fHotStandard) {
            return <p>{infoMap.veryHotInfo}</p>;
        } else {
            return <p>{infoMap.howInfo}</p>;
        }
    }
    class Temperature extends React.Component {
        constructor(props) {
            super(props);
            this.handleTempChange = this.handleTempChange.bind(this);
        }
        handleTempChange(e) {
            this.props.onTemperatureChange(e.target.value);
        }
        render() {
            const temperature = this.props.temperature;
```

```
        const type = this.props.type;
        return (
            <div>
                <span>
                    <label>{infoMap.outputInfo}{tempType[type]}{infoMap.outputInfoEnd}</label>
                    <input
                        name={type}
                        value={temperature}
                        onChange={this.handleTempChange} />
                    <TempLevel
                        wlevel={temperature}
                        type={type} />
                </span>
            </div>
        );
    }
}
class TemperatureApp extends React.Component {
    constructor(props) {
        super(props);
        this.state = {
            temperature: ''
        };
        this.handleCelsiusChange = this.handleCelsiusChange.bind(this);
        this.handleFahrenheitChange = this.handleFahrenheitChange.bind(this);
    }
    handleCelsiusChange(temperature) {
        this.setState({
            type: 'c',
            temperature
        });
    }
    handleFahrenheitChange(temperature) {
        this.setState({
            type: 'f',
            temperature
        });
    }
    render() {
        const temperature = this.state.temperature;
        return (
            <div>
                <Temperature
                    type="c"
                    temperature={temperature}
                    onTemperatureChange={this.handleCelsiusChange} />
                <Temperature
                    type="f"
```

```jsx
                    temperature={temperature}
                    onTemperatureChange={this.handleFahrenheitChange} />
            </div>
        );
    }
}
function toCelsius(fahrenheit) {
    return Math.round((fahrenheit - 32) * 5 / 9);
}
function toFahrenheit(celsius) {
    return Math.round((celsius * 9 / 5) + 32);
}
function toConvert(temperature, convert) {
    if (Number.isNaN(temperature)) {
        return '';
    }
    return convert(temperature).toString();
}
class TemperatureApp2 extends React.Component {
    constructor(props) {
        super(props);
        this.state = {
            temperature: '0',
            type: 'c'
        };
        this.handleCelsiusChange = this.handleCelsiusChange.bind(this);
        this.handleFahrenheitChange = this.handleFahrenheitChange.bind(this);
    }
    handleCelsiusChange(temperature) {
        this.setState({type: 'c', temperature});
    }
    handleFahrenheitChange(temperature) {
        this.setState({type: 'f', temperature});
    }
    render() {
        const type = this.state.type;
        const temperature = this.state.temperature;
        const celsius = type === 'f' ? toConvert(temperature, toCelsius) : temperature;
        const fahrenheit = type === 'c' ? toConvert(temperature, toFahrenheit) : temperature;
        return (
            <div>
                <Temperature
                    type="c"
                    temperature={celsius}
                    onTemperatureChange={this.handleCelsiusChange} />
                <Temperature
                    type="f"
```

```
                    temperature={fahrenheit}
                    onTemperatureChange={this.handleFahrenheitChange} />
            </div>
        );
    }
}
const exampleState= (
    <span>
        <h2 align="center">{infoMap.title1info}</h2>
        <hr/>
        <h3>{infoMap.title2info}</h3>
        <TemperatureApp />
        <hr/>
        <h3>{infoMap.title3info}</h3>
        <TemperatureApp2 />
    </span>
);
ReactDOM.createRoot(divReact).render(exampleState);
```

5.3.2 运行效果

运行文件 advancedstateexample.html，效果如图 5-4 所示。

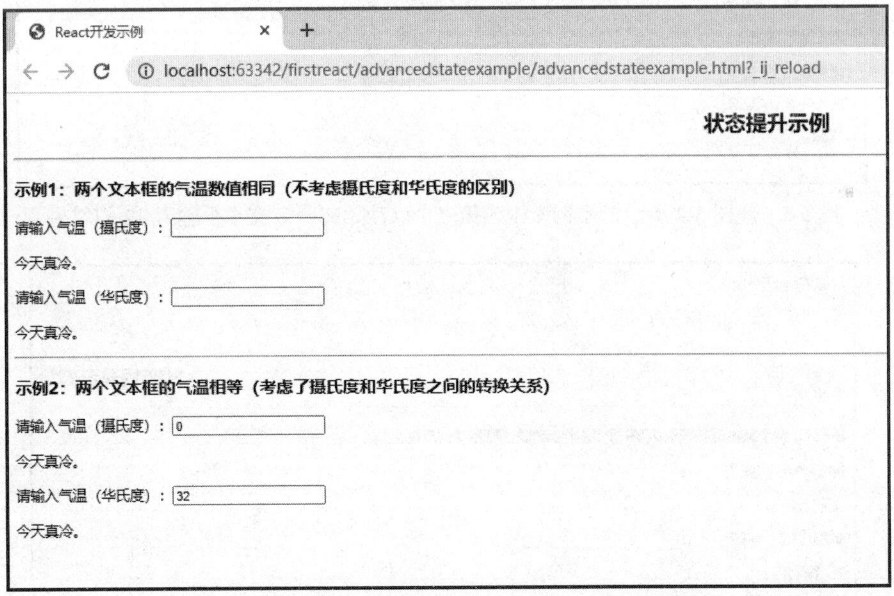

图 5-4 运行文件 advancedstateexample.html 的效果

在图 5-4 所示的示例 1 的第一个文本框（纵向从上向下）中输入数值（如 12），示例 1 的第二个文本框中的内容也变成相同的数值（如 12），效果如图 5-5 所示。可以发现第一个文本框和第二个文本框的数值同步变化（数值相同），而没有考虑摄氏度、华氏度之间的转换关系，于是图 5-5 所示中的示例 1 的第一个文本框对应的输出结果为"今天气温舒适。"，而第二个文本框对应的输出结果为"今天真冷。"。在图 5-4 所示的示例 1 中的第二个文本框(纵向从上向下)输入数值(如 12),示例 1 中的第一个文本框中的内容也变成相同的数值(如 12),效果如图 5-5 所示。

在图 5-4 所示的示例 2 中的第一个文本框(纵向从上向下)输入数值(如 45),示例 2 的第二个文本框中的内容变成输入摄氏度数值(如 45)对应的华氏度数值(如 113),效果如图 5-6 所示。可以发现第一个文本框和第二个文本框的数值同步变化(将摄氏度数值转换华氏度数值之后的结果),于是图 5-6 所示中的示例 2 的第一个文本框和第二个文本框对应的输出结果都是"今天真热。"。在图 5-4 所示的示例 2 的第二个文本框(纵向从上向下)中输入华氏度数值(如 113),示例 2 中的第一个文本框内容也变成对应的摄氏度数值(如 45),效果如图 5-6 所示。

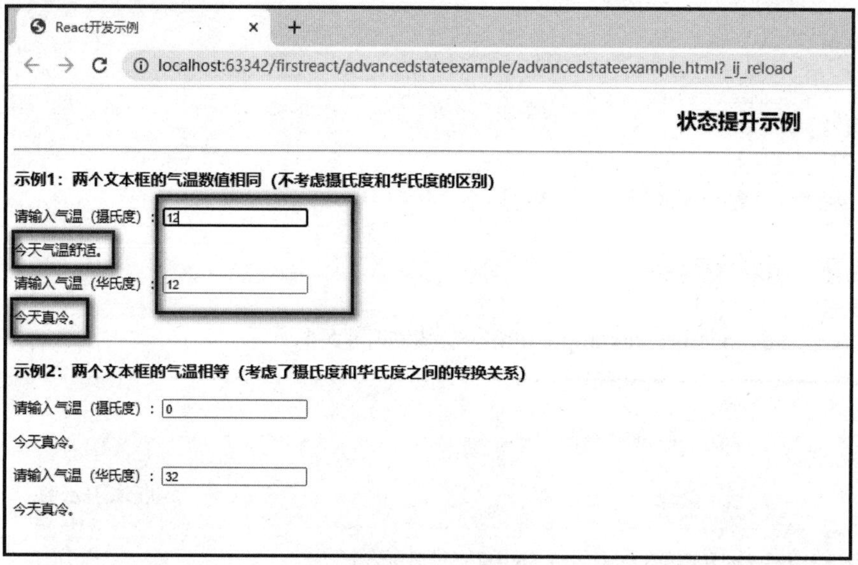

图 5-5 在图 5-4 所示的示例 1 中的第一个(或第二个)文本框输入 12 的效果

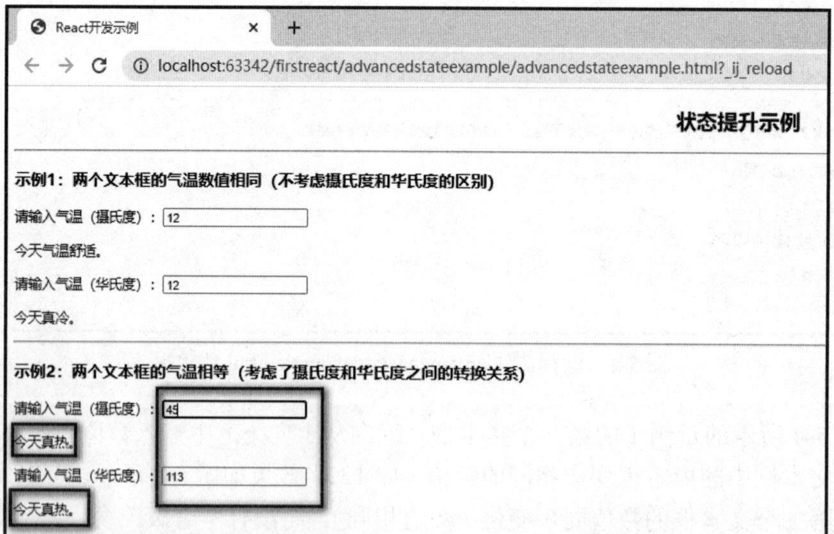

图 5-6 在图 5-4 所示的示例 2 的第一个文本框中输入 45(或在第二个文本框输入 113)的效果

习题 5

一、简答题

1. 简述对 state 的理解。
2. 简述对 setState()方法的理解。
3. 简述对 forceUpdate()方法的理解。
4. 简述对状态提升的理解。

二、实验题

1. 实现状态的基础应用开发。
2. 实现状态的提升应用开发。

第 6 章

React 表单

本章先简要介绍 React 表单、受控组件和非受控组件，再介绍表单组件和 ref 等内容。

6.1 React 表单概述

6.1.1 表单

在 React 里，HTML 表单元素和其他 DOM 元素有些不同。在 HTML 中，如<input>、<textarea>和<select>等表单元素（标签）通常自己维护 state（或称为数据、内容、文本等），并根据用户输入的内容进行更新。HTML 表单元素通常会保持一些内部 state（保存在组件的 state 属性中），并能用 setState()方法来更新其值。

在大多数情况下，使用 JavaScript 函数可以很方便地处理表单数据的提交等操作，同时还可以访问用户填写的表单数据。实现这种效果的标准方式是使用受控组件。在 React 受控组件中，state 成为唯一数据源，渲染表单的 React 组件控制用户在输入过程中对表单的操作。被 React 以这种方式控制取值的表单输入元素就叫作受控组件。

对于受控组件来说，输入的值始终由 React 的 state 驱动，也可以将 value（值）传递给其他 UI 界面元素，或者通过其他事件处理函数重置，但这意味着需要编写更多的代码。

6.1.2 受控组件

在 HTML 中，<textarea>元素通过其子元素定义其文本。而在 React 中，<textarea>使用 value 属性定义其 state。这使得使用<textarea>的表单和使用<input>的工作原理类似。

在 HTML 中，<select>是创建下拉列表标签，其 selected 属性对应的选项默认被选中。React 并不会使用 selected 属性，而是在根<select>标签上使用 value 属性。可以将数组传递到 value 属性中，以支持在<select>标签中选择多个选项。

总体来说，React 中 <input type="text">、<textarea> 和 <select> 之类的标签都非常相似。它们都接受一个 value 属性，可以使用它来实现受控组件。在受控组件上指定 value 属性的 props 会阻止用户更改输入。若指定了 value，但输入仍可编辑，则可能是因为将 value 设置为 undefined 或 null 的缘故。

当需要处理多个 input 元素时，可以给每个元素添加 name 属性，并让处理函数根据 event.target.name 的值选择要执行的操作具体元素。

6.1.3 非受控组件

在 React 中构建表单组件有好几种模式可用。在大多数情况下，可以使用受控组件来处理表单数据。在受控组件中，表单数据是由 React 组件来管理的。有时使用受控组件会比较麻烦，因为需要为数据更新（变化）的每种方式都编写事件处理函数，并通过一个 React 组件传递所有的输入 state。在这些情况下，可能希望使用非受控组件来实现输入表单。

在非受控组件中，表单数据将交由 DOM 节点来处理。在编写非受控组件时，可以使用 ref 从 DOM 节点中获取表单数据。在 React 中，ref 是一个对象，存储着一个组件整个生命周期内的值，可以使用 ref 特性直接访问 DOM 节点。

在 HTML 中，<input type="file"> 允许用户从存储设备中选择一个或多个文件，将其上传到服务器，或通过使用 JavaScript 的文件 API 进行控制。因为它的 value 是只读的，所以它是 React 中的非受控组件。在 React 中，<input type="file" /> 始终是一个非受控组件，因为它的值只能由用户设置，而不能通过代码控制。应该使用文件 API 与文件进行交互。

因为非受控组件将真实数据储存在 DOM 节点中，所以有时候使用非受控组件较容易集成 React 和非 React 代码。如果不介意代码是否美观，并且希望快速编写代码，使用非受控组件往往可以减少代码量。否则，应使用受控组件。

在 React 渲染生命周期时，表单元素上的 value 将覆盖 DOM 节点中的值，在非受控组件中，用户经常希望 React 能赋予组件一个初始值，但不希望其能控制后续的更新。在这种情况下，可以指定一个 defaultValue 属性，而不是 value。<input type="checkbox"> 和 <input type="radio"> 支持 defaultChecked，<select> 和 <textarea> 支持 defaultValue。

6.2 表单组件

6.2.1 开发示例

在项目 firstreact 根目录下创建 formexample 子目录，在 firstreact\formexample 目录下创建文件 formexample.html，代码与例 1-3 所示的代码相同。在 firstreact\formexample 目录下创建文件 index.js，代码如例 6-1 所示。

【例 6-1】 在 firstreact\formexample 目录下创建的文件 index.js 的代码。

```
const divReact = document.getElementById('root');
const infoMap= {
    namezsf:'张三丰',
```

```
        nameInfo:'姓名: ',
        namechangeInfo:'用户名变更成"',
        namechangeInfoEnd:'"。',
        submitInfo:'提交',
        namechangeInfoBegin:'"',
}
//使用state,没有绑定按钮事件处理功能
class FormComp extends React.Component {
    constructor(props) {
        super(props);
        this.state = {username: infoMap.namezsf};
    }
    render() {
        return (
            <form>
                <label>{infoMap.nameInfo}</label>
                <input type="text" name="name" value={this.state.username} />
                <input type="submit" value={infoMap.submitInfo} />
            </form>
        );
    }
}
//绑定按钮事件
class FormComp2 extends React.Component {
    constructor(props) {
        super(props);
        this.state = {username: infoMap.namezsf};
        this.handleChange = this.handleChange.bind(this);
    }
    handleChange(event) {
        let eUsername = event.target.value;
        this.setState({username: eUsername});
        alert(infoMap.namechangeInfo + eUsername + infoMap.namechangeInfoEnd);
    }
    render() {
        return (
            <form>
                <label>{infoMap.nameInfo}</label>
      <input type="text" name="name" value={this.state.username} onChange=
{this.handleChange} />
                <input type="submit" value={infoMap.submitInfo} />
            </form>
        );
    }
}
class FormComp3 extends React.Component {
    constructor(props) {
        super(props);
        this.state = {
```

```
            firstname: '',
        };
        this.handleFirstNameChange = this.handleFirstNameChange.bind(this);
    }
    handleFirstNameChange(event) {
        let targetValue = event.target.value;
        let targetValueUpper = targetValue.toUpperCase();
        this.setState({firstname: targetValueUpper});
        alert(infoMap.namechangeInfoBegin+targetValue+
            infoMap.namechangeInfoBegin+ infoMap.namechangeInfo +
            targetValueUpper + infoMap.namechangeInfoEnd);
    }
    render() {
        return (
            <form>
                <input type="text" name="firstname"
                    value={this.state.firstname}
                    onChange={this.handleFirstNameChange} />
            </form>
        );
    }
}
class FormComp4 extends React.Component {
    constructor(props) {
        super(props);
        this.state = {
            tel: '',
            info: '13800138000'
        };
        this.handleTelChange = this.handleTelChange.bind(this);
    }
    handleTelChange(event) {
        //用正则表达式表示正确的手机号
        let telReg = /13[0-9]\d{8}|18[56789]\d{8}/g;
        let targetTel = event.target.value;
        let finalTel = targetTel;
        if(targetTel.length < 11) {
            this.setState({info: 'pls go on...'});
        } else if(targetTel.length == 11) {
            if(telReg.test(targetTel)) {
                this.setState({info: 'right number.'});
                finalTel = targetTel;
            } else {
                this.setState({info: 'wrong number.'});
                finalTel = "";
            }
        } else {
            finalTel = targetTel.substr(0, 11);
        }
```

```jsx
                this.setState({tel: finalTel});
            }
            render() {
                return (
                    <form>
                        <input type="text" name="tel"
                            value={this.state.tel}
                            onChange={this.handleTelChange} />
                        <input type="text" name="info" className="css-input-info"
                            value={this.state.info} readOnly/>
                        <input type="submit" value="重置" />
                    </form>
                );
            }
        }
        class FormComp5 extends React.Component {
            constructor(props) {
                super(props);
                this.state = {
                    sn: ''
                };
                this.handleSNChange = this.handleSNChange.bind(this);
            }
            handleSNChange(event) {
                let targetSN = event.target.value;
                let finalSN = targetSN;
                if(targetSN.length > 11) {
                    finalSN = targetSN.substr(0, 11);
                } else {
                    if(targetSN.length == 3) {
                        finalSN = targetSN + "-";
                    } else if(targetSN.length == 7) {
                        finalSN = targetSN + "-";
                    } else {}
                }
                this.setState({sn: finalSN});
            }
            render() {
                return (
                    <form>
                        <input type="text" name="sn"
                            value={this.state.sn}
                            onChange={this.handleSNChange} />
                        <input type="submit" value="重置" />
                    </form>
                );
            }
        }
        class FormComp6 extends React.Component {
```

```
        constructor(props) {
            super(props);
            this.state = {
                keyword: ''
            };
            this.handleArticleChange = this.handleArticleChange.bind(this);
        }
        handleArticleChange(event) {
            let regexp_keyword = ^sking\s/g;
            let target_keyword = event.target.value;
            let markdown_keyword = target_keyword;
            if(regexp_keyword.test(target_keyword)) {
                markdown_keyword=target_keyword.replace(regexp_keyword,"KING");
            }
            this.setState({keyword: markdown_keyword});
        }
        render() {
            return (
                <form>
                    <textarea type="text" name="article"
                            value={this.state.keyword}
                            onChange={this.handleArticleChange} />
                    <input type="submit" value="提交" />
                </form>
            );
        }
    }
    class FormComp7 extends React.Component {
        constructor(props) {
            super(props);
            this.state = {
                selval: 'football'
            };
            this.handleSelChange = this.handleSelChange.bind(this);
        }
        handleSelChange(event) {
            let selVal = event.target.value;
            console.log("You has selected '" + selVal + "'.");
            this.setState({
                selval: selVal
            });
        }
        render() {
            return (
                <form>
                    <label>
                        请选择您喜欢的运动:
                        <select value={this.state.selval} onChange=
{this.handleSelChange}>
```

```jsx
                    <option value="baseball">Baseball</option>
                    <option value="football">Football</option>
                    <option value="basketball">Basketball</option>
                </select>
            </label>
            <input type="submit" value="提交" />
        </form>
    );
  }
}
class FormComp8 extends React.Component {
    constructor(props) {
        super(props);
        this.state = {
            isOnOff: true,
            username: "king"
        };
        this.handleInputChange = this.handleInputChange.bind(this);
    }
    handleInputChange(event) {
        const target = event.target;
        const value = target.type === 'checkbox'?target.checked : target.value;
        const name = target.name;
        this.setState({
            [name]: value
        });
        console.log(name + " : " + value);
    }
    render() {
        return (
            <form>
                <label>
                    On/Off:  
                    <input
                        name="isOnOff"
                        type="checkbox"
                        checked={this.state.isOnOff}
                        onChange={this.handleInputChange} />
                </label>
                <br/>
                <label>
                    Username:  
                    <input
                        name="username"
                        type="text"
                        value={this.state.username}
                        onChange={this.handleInputChange} />
                </label>
```

```
            </form>
        );
    }
}
class FormComp9 extends React.Component {
    constructor(props) {
        super(props);
        this.state = {
            username: "",
            age: "18",
            gender: "male"
        };
        this.handleInputChange = this.handleInputChange.bind(this);
        this.handleSelChange = this.handleSelChange.bind(this);
        this.handleSubmit = this.handleSubmit.bind(this);
    }
    handleInputChange(event) {
        const target = event.target;
        const value = target.type === 'checkbox' ? target.checked : target.value;
        const name = target.name;
        this.setState({
            [name]: value
        });
    }
    handleSelChange(event) {
        let gender = event.target.value;
        this.setState({
            gender: gender
        });
    }
    handleSubmit(event) {
        event.preventDefault();
        console.log("Username : " + this.state.username);
        console.log("Age : " + this.state.age);
        console.log("Gender : " + this.state.gender);
    }
    render() {
        return (
            <form onSubmit={this.handleSubmit}>
                <label>
                    Username:  
                    <input
                        name="username"
                        type="text"
                        value={this.state.username}
                        onChange={this.handleInputChange} />
                </label><br/>
                <label>
```

```
                        Age:  
                        <input
                            name="age"
                            type="number"
                            value={this.state.age}
                            onChange={this.handleInputChange} />
                    </label><br/>
                    <label>
                        Gender:  
                <select name="gender" value={this.state.gender} onChange=
{this.handleSelChange}>
                            <option value="male">Male</option>
                            <option value="female">Female</option>
                        </select>
                    </label><br/><br/>
                    <input type="submit" value="提交" />
                </form>
            );
        }
    }
    const exampleForm = (
        <span>
            <h2 align="center">Form 简单示例</h2>
            <hr/>
            <div>使用 state</div> <FormComp />
            <hr/>
            <div>组件</div> <FormComp2 />
            <hr/>
            <div>字母大写（小写会变成大写）</div><FormComp3 />
             <hr/>
            <div>手机号(+86)</div><FormComp4 />
            <hr/>
            <div>格式化序列号</div><FormComp5 />
            <hr/>
            <div>多行输入（Textare）</div><FormComp6 />
            <hr/>
            <div>选择（Select）标签</div><FormComp7 />
            <hr/>
            <div>Multi Input 标签</div><FormComp8 />
            <hr/>
            <div>Submit 标签</div><FormComp8 />
        </span>
    );
    ReactDOM.createRoot(divReact).render(exampleForm);
```

6.2.2 运行效果

运行文件 formexample.html，效果如图 6-1 所示。对图 6-1 所示中的文本框、按钮、选择等标签的操作，请读者根据提供的源代码自己完成操作或参考视频的操作来观察效果。

图 6-1　运行文件 formexample.html 的效果

视频讲解

6.3　ref

　　ref 提供了一种用于访问 DOM 节点或 React 元素的方法。可以简单地将 ref 理解成 DOM 节点或 React 元素的代理。其中，DOM 节点（从原生的角度来表述）和 React 元素（从 React 的角度来表述）从本质上看是等价的。当 ref 属性用于创建 HTML 元素的场景时，在构造函数中使用 React.createRef()方法创建的 ref 能够接收底层 DOM 元素，并将其作为它的 current 属性。当 ref 属性用于定义自定义类组件（是一种 React 元素）的场景时，ref 对象可以接收类组件的实例，将其作为它的 current 属性。此外，还可以在函数组件内部使用 ref 属性，并将它指向一个 DOM 元素或类组件（DOM 节点）。焦点管理、文本选择、媒体播放、触发强制动画、集成第三方 DOM 库等情况下，适合使用 ref。

6.3.1　开发示例

　　在项目 firstreact 根目录下创建 formrefsexample 子目录，在 firstreact\formrefsexample 目录下创建文件 formrefsexample.html，代码与例 1-3 所示的代码相同。在 firstreact\

formrefsexample 目录下创建文件 index.js，代码如例 6-2 所示。

【例 6-2】 在 firstreact\formrefsexample 目录下创建的文件 index.js 的代码。

```
const divReact = document.getElementById('root');
const infoMap= {
    usernameInfo:'用户名为：',
    title1Info:'Refs - 受控组件',
    inputlabelInfo:'请输入用户名  ',
    spaceInfo:'  ',
    title2Info:'Refs - 非受控组件',
    title3Info:'Refs 简单示例',
    submitinfo:'提交',
}
class ReactRefsComp extends React.Component {
    constructor(props) {
        super(props);
        this.state = {
            username: '',
            output: ''
        };
        this.handleChange = this.handleChange.bind(this);
        this.handleSubmit = this.handleSubmit.bind(this);
    }
    handleChange(e) {
        this.setState({
            username: e.target.value
        });
    }
    handleSubmit(e) {
        e.preventDefault();
        this.setState({
            output: infoMap.usernameInfo + this.state.username
        });
    }
    render() {
        return (
            <form onSubmit={this.handleSubmit}>
                <h3>{infoMap.title1Info}</h3>
                <label>{infoMap.inputlabelInfo}
                    <input
                        type="text"
                        value={this.state.username}
                        onChange={this.handleChange}
                    />
                </label>{infoMap.spaceInfo}
                <input type="submit" value={infoMap.submitinfo} />
                <br/><br/>
                <p>{this.state.output}</p>
            </form>
```

```
            );
        }
    }
    class ReactRefsComp2 extends React.Component {
        constructor(props) {
            super(props);
            this.state = {
                output: ''
            };
            this.handleSubmit = this.handleSubmit.bind(this);
        }
        handleSubmit(e) {
            e.preventDefault();
            this.setState({
                output: infoMap.usernameInfo + this.username.value
            });
        }
        render() {
            return (
                <form onSubmit={this.handleSubmit}>
                    <h3>{infoMap.title2Info}</h3>
                    <label>{infoMap.inputlabelInfo}
                        <input
                            type="text"
                            ref={input => {this.username = input}}
                        />
                    </label>{infoMap.spaceInfo}
                    <input type="submit" value={infoMap.submitinfo} />
                    <br/><br/>
                    <p>{this.state.output}</p>
                </form>
            );
        }
    }
    const exampleRefs = (
        <span>
            <h3 align="center">{infoMap.title3Info}</h3>
            <hr/>
            <ReactRefsComp/>
            <hr/>
            <ReactRefsComp2/>
        </span>
    );
    ReactDOM.createRoot(divReact).render(exampleRefs);
```

6.3.2 运行效果

运行文件 formrefsexample.html，效果如图 6-2 所示。

在图 6-2 所示中示例的第一个文本框（纵向从上往下）中输入用户名（如 zs），效果如图 6-3 所示。在图 6-3 所示中示例的第二个文本框中输入用户名（如"李斯"），效果如图 6-4 所示。

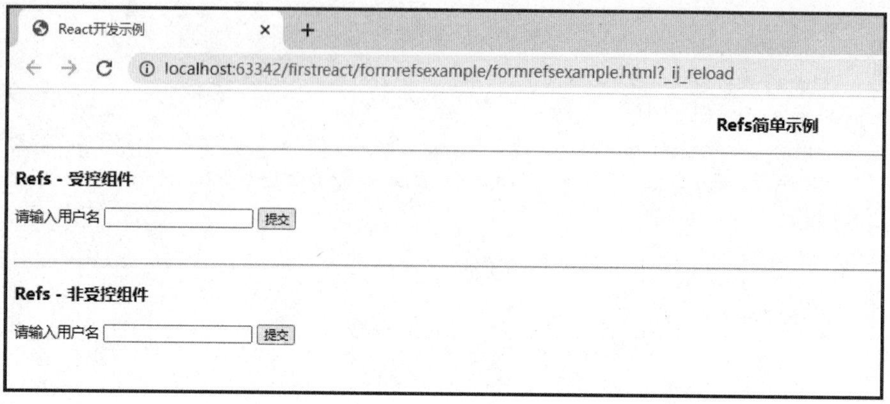

图 6-2 运行文件 formrefsexample.html 的效果

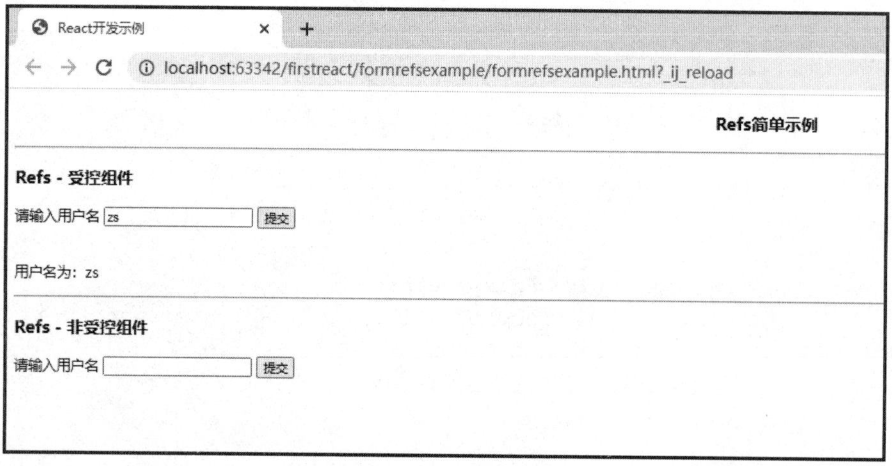

图 6-3 在图 6-2 所示中示例的第一个文本框中输入 zs 的效果

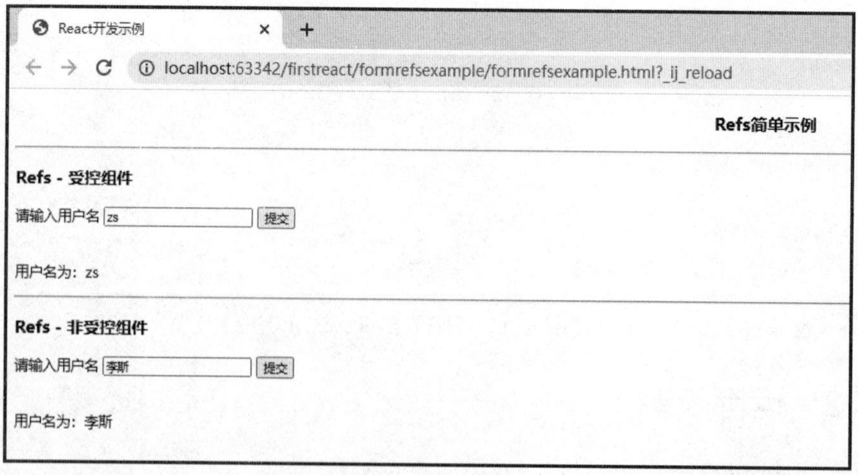

图 6-4 在图 6-3 所示中示例的第二个文本框中输入"李斯"的效果

习题 6

一、简答题

1. 简述对表单的理解。
2. 简述对受控组件的理解。
3. 简述对非受控组件的理解。

二、实验题

1. 完成表单组件的应用开发。
2. 完成 ref 的应用开发。

第 7 章

React组件的组合和继承

本章先简要介绍组件的组合和继承，再介绍带样式的组合组件、页面布局、特例关系组合和类组合等内容。

7.1 React 组件的组合和继承概述

7.1.1 组合

如第 2 章介绍的那样，可以在开发小组件的基础上，通过组件组合的方式实现更大的组件。React 有十分强大的组件组合模式。React 官方推荐使用组合而非继承来实现组件间的代码复用。通过分析发现，在 Facebook 中有成百上千个组件中使用到了 React，但是并没有发现需要使用继承来构建组件层次的情况。

React 中可以将数据、函数等内容作为 props 进行传递。特殊组件可以通过 props 定制并渲染一般组件。有时，可以不使用 children，而是在一个组件中预留出几个空的位置，自行约定将所需内容传入 props 并使用相应的 props。props 和组合提供了清晰而安全的定制组件外观和行为的灵活方式。组件可以接收任意 props，包括基本数据类型、React 元素以及函数。

7.1.2 继承

如第 2 章介绍的那样，还可以使用类来自定义组件，即类组件。定义类组件时，类组件要继承 React.Component。

有时，可以把一些组件看作是其他组件的特殊实例，即存在一种继承关系。例如 WelcomeDialog 可以说是 Dialog 的特殊实例。

如果想要在组件间复用非 UI 界面的功能，可以将其提取为一个单独的 JavaScript 模块,如函数、对象或者类。组件可以直接引入(import)而无须通过扩展(extend)来继承它们。

7.2 带样式的组合组件

7.2.1 引入包、样式和功能文件

在项目 firstreact 根目录下创建 composedexample 子目录,在 firstreact\composedexample 目录下创建文件 composedexample.html,代码如例 7-1 所示。例 7-1 所示的代码与例 1-3 所示的代码的差别在于增加了一行语句(不考虑注释代码)来导入样式表文件 index.css。

【例 7-1】 在 firstreact\composedexample 目录下创建的文件 composedexample.html 的代码。

```
<!DOCTYPE html>
<html lang="en">
<head>
    <meta charset="UTF-8">
    <title>React 开发示例</title>
    <script src="https://unpkg.com/react@18.2.0/umd/react.production.min.js"></script>
    <script src="https://unpkg.com/react-dom@18.2.0/umd/react-dom.production.min.js"></script>
    <script src="https://unpkg.com/@babel/standalone@7.18.0/babel.min.js"></script>
    <!--新增下一行代码导入样式文件 index.css-->
    <link href="index.css" rel="stylesheet" type="text/css"/>
</head>
<body>
<div id="root"></div>
<script type="text/babel" src="index.js"></script>
</body>
</html>
```

7.2.2 定义样式

在 firstreact\composedexample 目录下创建文件 index.css,代码如例 7-2 所示。

【例 7-2】 在 firstreact\composedexample 目录下创建的文件 index.css 的代码。

```
body {
    text-align: center;
}
div#root {
    position: relative;
    width: 400px;
    height: auto;
```

```css
    border: 0px solid blue;
    margin: 8px auto;
    padding: 8px;
    font: normal 16px/1.6em "Microsoft Yahei", arial, sans-serif;
    text-align: center;
}
div#root .dialogBorder-red {
    border-width: 2px 8px 8px 2px;
    border-style: solid;
    border-color: red;
    margin: 8px auto;
    padding: 8px;
}
div#root .dialog-title {
    text-align: left;
    color: blue;
}
div#root .dialog-message {
    text-align: center;
    color:green;
}
div#root .dialog-info {
    font-style: italic;
    text-align: left;
    color:yellow;
}
```

7.2.3 定义功能

在 firstreact\composedexample 目录下创建文件 index.js,代码如例 7-3 所示。

【例 7-3】 在 firstreact\composedexample 目录下创建的文件 index.js 的代码。

```js
const divReact = document.getElementById('root');
const infoMap= {
    welcomeInfo:'欢迎',
    greetInfo:'谢谢访问!',
    attentionInfo:'提示信息',
    detailInfo:'请阅读详细内容。',
    welcomeDialogInfo:'自定义 Dialog 类型: WelcomeDialog。',
    infoDialogInfo:'"自定义 Dialog 类型: InfoDialog。',
}
function MyDialogBox(props) {
    return (
        <div className={'dialogBorder-' + props.color}>
            {props.children}
        </div>
    );
}
```

```jsx
function WelcomeDialog() {
    return (
        <MyDialogBox color="red">
            <h3 className="dialog-title">
                {infoMap.welcomeInfo}
            </h3>
            <p className="dialog-message">
                {infoMap.greetInfo}
            </p>
        </MyDialogBox>
    );
}
function InfoDialog() {
    return (
        <MyDialogBox color="red">
            <h3 className="dialog-title">
                {infoMap.attentionInfo}
            </h3>
            <p className="dialog-info">
                {infoMap.detailInfo}
            </p>
        </MyDialogBox>
    );
}
function MyDialog(props) {
    const dType = props.dlgType;
    if(dType) {
        alert(infoMap.welcomeDialogInfo);
        return <WelcomeDialog />;
    } else {
        alert(infoMap.infoDialogInfo);
        return <InfoDialog />;
    }
}
const reactDialog = (
    <span>
    <MyDialog dlgType={true}/>
    </span>
);
ReactDOM.createRoot(divReact).render(reactDialog);
```

7.2.4 带样式组件综合应用的运行效果

运行文件 composedexample.html，效果如图 7-1 所示。单击图 7-1 所示中的"确定"按钮，效果如图 7-2 所示。

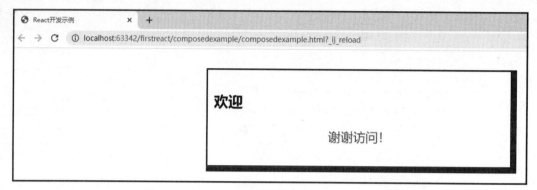

图 7-1　运行文件 composedexample.html 的效果

图 7-2　单击图 7-1 所示中的"确定"按钮的效果

视频讲解

7.3　页面布局

7.3.1　定义样式和功能

在项目 firstreact 根目录下创建 pagestructexample 子目录，在 firstreact\pagestructexample 目录下创建文件 pagestructexample.html，代码与例 7-1 所示的代码相同。在 firstreact\pagestructexample 目录下创建文件 index.css，代码如例 7-4 所示。

【例 7-4】　在 firstreact\ pagestructexample 目录下创建的文件 index.css 的代码。

```
body {
    text-align: center;
}
div#root {
    position: relative;
    width: 400px;
    height: auto;
    border: 0px solid black;
    margin: 8px auto;
    padding: 8px;
    font: normal 16px/1.6em "Microsoft Yahei", arial, sans-serif;
    text-align: center;
}
div#root .main-panel-header {
    margin: 2px auto;
```

```css
    padding: 2px;
    width: 100%;
    height: 32px;
    text-align: center;
}
div#root .main-panel-left {
    float: left;
    margin: 2px auto;
    padding: 2px;
    width: 33%;
    height: auto;
    text-align: center;
}
div#root .main-panel-right {
    float: left;
    margin: 2px auto;
    padding: 2px;
    width: 65%;
    height: auto;
    text-align: center;
}
div#root .css-header {
    font-weight: bold;
    border: 1px solid ;
    background-color: #aaa;
    color: red;
}
div#root .css-left {
    height: 96px;
    border: 1px solid red;
    background-color: #ccc;
    color: red;
}
div#root .css-right {
    height: 96px;
    border: 1px solid red;
    background-color: #eee;
    color: red;
}
```

在 firstreact\pagestructexample 目录下创建文件 index.js，代码如例 7-5 所示。

【例 7-5】 在 firstreact\pagestructexample 目录下创建的文件 index.js 的代码。

```js
const divReact = document.getElementById('root');
const infoMap= {
    pageInfo:'页面头部',
    underLeftInfo:'页面左下部分',
    underRightInfo:'页面右下部分',
}
```

```
function HeaderPanel() {
    return <div className="css-header">{infoMap.pageInfo}</div>;
}
function LeftPanel() {
    return <div className="css-left">{infoMap.underLeftInfo}</div>;
}
function RightPanel() {
    return <div className="css-right">{infoMap.underRightInfo}</div>;
}
function MainPanel(props) {
    return (
        <div className="main-panel">
            <div className="main-panel-header">
                {props.header}
            </div>
            <div className="main-panel-left">
                {props.left}
            </div>
            <div className="main-panel-right">
                {props.right}
            </div>
        </div>
    );
}
function Page() {
    return (
        <MainPanel
            header={
                <HeaderPanel />
            }
            left={
                <LeftPanel />
            }
            right={
                <RightPanel />
            } />
    );
}
const pageDialog = (
    <span>
    <Page/>
    </span>
);
ReactDOM.createRoot(divReact).render(pageDialog);
```

7.3.2 运行效果

运行文件 pagestructexample.html，效果如图 7-3 所示。

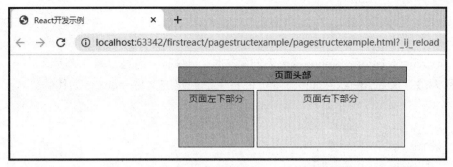

图 7-3 运行文件 pagestructexample.html 的效果

7.4 特例关系组合

7.4.1 定义样式和功能

在 firstreact 根目录下创建 relationexample 子目录，在 firstreact\relationexample 目录下创建文件 relationexample.html，代码与例 7-1 所示的代码相同。在 firstreact\ relationexample 目录下创建文件 index.css，代码如例 7-6 所示。

【例 7-6】 在 firstreact\relationexample 目录下创建的文件 index.css 的代码。

```css
body {
    text-align: center;
}
div#root {
    position: relative;
    width: 400px;
    height: auto;
    border: 0px solid black;
    margin: 8px auto;
    padding: 8px;
    font: normal 16px/1.6em "Microsoft Yahei", arial, sans-serif;
    text-align: center;
}
div#root .dialogBorder-gray {
    border-width: 2px 8px 8px 2px;
    border-style: solid;
    border-color: red;
    margin: 8px auto;
    padding: 8px;
}
div#root  .dialog-title {
    text-align: left;
    color:black;
}
div#root  .dialog-message {
    text-align: center;
```

```
        color:blue;
}
```

在 firstreact\relationexample 目录下创建文件 index.js，代码如例 7-7 所示。

【例 7-7】 在 firstreact\relationexample 目录下创建的文件 index.js 的代码。

```
const divReact = document.getElementById('root');
const infoMap= {
    welcomeInfo:'欢迎',
    greetInfo:'谢谢访问！',
}
function DialogBox(props) {
    return (
        <div className={'dialogBorder-' + props.color}>
            {props.children}
        </div>
    );
}
function Dialog(props) {
    return (
        <DialogBox color="gray">
            <h3 className="dialog-title">
                {props.title}
            </h3>
            <p className="dialog-message">
                {props.message}
            </p>
        </DialogBox>
    );
}
function WelcomeDialog() {
    return (
        <Dialog
            title={infoMap.welcomeInfo}
            message={infoMap.greetInfo} />
    );
}
const reactDialog = (
    <span>
    <WelcomeDialog/>
    </span>
);
ReactDOM.createRoot(divReact).render(reactDialog);
```

7.4.2 运行效果

运行文件 relationexample.html，效果如图 7-4 所示。

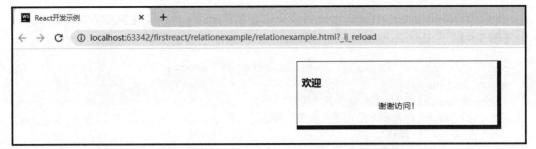

图 7-4 运行文件 relationexample.html 的效果

视频讲解

7.5 类组合

7.5.1 定义样式和功能

在项目 firstreact 根目录下创建 classcomposeexample 子目录，在 firstreact\classcomposeexample 目录下创建文件 classcomposeexample.html，代码与例 7-1 所示的代码相同。在 firstreact\classcomposeexample 目录下创建文件 index.css，代码如例 7-8 所示。

【例 7-8】 在 firstreact\classcomposeexample 目录下创建的文件 index.css 的代码。

```css
body {
    text-align: center;
}
div#root {
    position: relative;
    width: 400px;
    height: auto;
    border: 0px solid black;
    margin: 8px auto;
    padding: 8px;
    font: normal 16px/1.6em "Microsoft Yahei", arial, sans-serif;
    text-align: center;
}
div#root .dialogBorder-gray {
    border-width: 2px 8px 8px 2px;
    border-style: solid;
    border-color: #ccc;
    margin: 8px auto;
    padding: 8px;
}
div#root .dialog-title {
    text-align: left;
}
div#root .dialog-message {
    text-align: center;
}
```

在 firstreact\classcomposeexample 目录下创建文件 index.js，代码如例 7-9 所示。

【例 7-9】 在 firstrcact\classcomposeexample 目录下创建的文件 index.js 的代码。

```
const divReact = document.getElementById('root');
const infoMap= {
    confirmInfo:'您刚刚单击了"确定"按钮。',
    cancelInfo:'您刚刚单击了"取消"按钮。',
    titleInfo:'确认',
    selectInfo:'请确认您的选择。',
    btnConfirmValue:'确定',
    btnCancelValue:'取消',
}
function DialogBox(props) {
    return (
        <div className={'dialogBorder-' + props.color}>
            {props.children}
        </div>
    );
}
function Dialog(props) {
    return (
        <DialogBox color="gray">
            <h3 className="dialog-title">
                {props.title}
            </h3>
            <p className="dialog-message">
                {props.message}
            </p>
            {props.children}
        </DialogBox>
    );
}
class ConfirmDialog extends React.Component {
    constructor(props) {
        super(props);
        this.handleConfirmClick = this.handleConfirmClick.bind(this);
        this.handleCancelClick = this.handleCancelClick.bind(this);
    }
    handleConfirmClick(event) {
        alert(infoMap.confirmInfo);
    }
    handleCancelClick(event) {
        alert(infoMap.cancelInfo);
    }
    render() {
        return (
            <Dialog
                title={infoMap.titleInfo}
                message={infoMap.selectInfo}>
```

```
                <button onClick={this.handleConfirmClick}>
{infoMap.btnConfirmValue}</button>
                <button onClick={this.handleCancelClick}>
{infoMap.btnCancelValue}</button>
            </Dialog>
        );
    }
}
const rDialog = (<span><ConfirmDialog/></span>);
ReactDOM.createRoot(divReact).render(rDialog);
```

7.5.2 运行效果

运行文件 classcomposeexample.html，效果如图 7-5 所示。单击图 7-5 所示中的"确定"按钮，效果如图 7-6 所示。单击图 7-6 所示中的"确定"按钮，效果如图 7-5 所示。单击图 7-5 所示中的"取消"按钮，效果如图 7-7 所示。单击图 7-7 所示中的"确定"按钮，效果如图 7-5 所示。

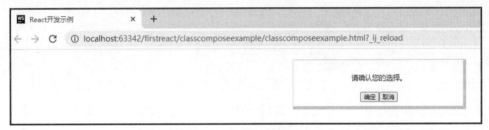

图 7-5　运行文件 classcomposeexample.html 的效果

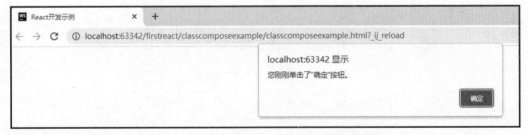

图 7-6　单击图 7-5 所示中的"确定"按钮的效果

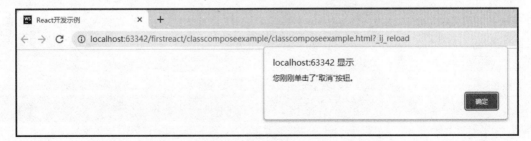

图 7-7　单击图 7-5 所示中的"取消"按钮的效果

习题 7

一、简答题

1. 简述对组件继承的理解。
2. 简述对组件组合的理解。

二、实验题

1. 完成带样式的组合组件的应用开发。
2. 完成页面布局的应用开发。
3. 完成特例关系组合的应用开发。
4. 完成类组合的应用开发。

第二部分 高阶篇

第 8 章　React 使用 Hook 增强组件
第 9 章　React 基础原理和高级指引
第 10 章　React 应用开发的工具

第 8 章

React使用Hook增强组件

本章先简要介绍 Hook、常见的 Hook API、自定义 Hook 和 Hook 使用规则，再介绍 State Hook、Effect Hook 和其他 Hook 的应用等内容。

8.1 Hook 概述

8.1.1 Hook

Hook（钩子）是从 React 16.8 开始新增加的特性。Hook 是一些可以在函数组件里钩入 props、state、context、ref 及生命周期等特性的函数。不能在类组件中使用 Hook。Hook 中的数据发生变化后，Hook 会使用新数据重新渲染所在的组件。

React 应用开发时，是否使用 Hook 是可选的，无须重写任何已有代码就可以在一些组件中使用 Hook，而且 Hook 是向后兼容的。在 Hook 之前，React 没有直接提供将可复用的内容附加到组件的方法。有一些替代方案，如高阶组件。但是这些替代方案需要重新组织组件结构，这可能会比较麻烦且使代码难以理解。

一般来说，组件起初很简单，但是逐渐充斥着状态逻辑和副作用（即出现退化现象）。生命周期方法中常常包含一些不相关的逻辑。相互关联且需要对照修改的代码被进行了拆分，而不相关的代码却组合在同一个方法中。这些代码很容易产生 bug，并且导致逻辑不一致。可以使用 Hook 从组件中提取状态逻辑，这些逻辑可以单独测试并复用。

在多数情况下，不可能将组件拆分为更小的粒度，因为状态逻辑无处不在。这给测试带来了一定的挑战。这也是将 React 与状态管理库（如 Redux、MobX、Recoil、XState）结合使用的原因之一。但是，将 React 与状态管理库结合使用会引入很多抽象概念，需要在不同的文件之间来回切换，使组件复用变得更加困难。为了解决这个问题，Hook 将组件中相互关联的部分拆分成更小的函数（如设置订阅或请求数据），而并非强制按照生命周期划分组件。还可以使用 reducer（缩减器、折叠器）来管理组件的内部状态，使其更加可预测。

React 官方表示 Hook 将会覆盖所有使用类组件的场景，但是也继续为类组件提供支持，可以在新的代码中同时使用 Hook 和类组件。

8.1.2　Hook API

React 内置了一些如 useState()函数和 useEffect()函数等的 Hook。为了和前面方法区分，在 Hook 中称其为函数。

（1）useState()函数返回一个 state 并更新 state。在初始渲染期间，返回的 state 与传入的第一个参数（initialState）值相同。在后续的重新渲染中，useState()函数返回的第一个值将始终是更新后的最新的 state。如果新的 state 需要通过使用先前的 state 计算得出，那么可以将 useState()函数传递给 setState()方法。与 setState()方法不同，useState()函数不会自动合并更新对象。

（2）useEffect()函数是接收包含命令式且可能有副作用（effect）代码的函数。在函数组件被 React 渲染时，不允许改变 DOM、添加订阅、设置定时器、记录日志以及执行其他包含副作用的操作，因为这可能会产生莫名其妙的 bug 并破坏 UI 界面的一致性。在默认情况下，effect 会在每轮渲染结束后执行，也可以选择让它在只有某些值改变的时候才执行。一旦 effect 的依赖发生变化，它就会被重新创建。通常，组件卸载时需要清除 effect 创建的诸如订阅或计时器 ID 等资源。要实现这一点，useEffect()函数需要返回一个清除函数。为防止内存泄漏，清除函数会在组件卸载前执行。

（3）useContext(MyContext)函数接收一个 context 对象（React.createContext()方法的返回值）并返回该 context 的当前值。利用 useContext(MyContext)函数能够读取 context 值、订阅 context 的变化。需要在上层组件树中使用 <MyContext.Provider> 来为下层组件提供 context。当前的 context 值由上层组件中距离当前组件最近的<MyContext.Provider>的 props 决定。当组件上层最近的 <MyContext.Provider> 更新时，useContext(MyContext)函数会触发重新渲染，并使用最新传递给 MyContext provider 的 context 值进行渲染。调用了 useContext (MyContext)函数的组件总会在 context 值变化时重新渲染。

（4）useReducer()函数是 useState()函数的替代方案。它接收一个形如(state, action) => newState 的 reducer，并返回当前的 state 以及与其配套的 dispatch()函数。在某些场景下，useReducer()函数会比 useState()函数更适用。例如，state 逻辑较复杂且包含多个子值，或者下一个 state 依赖于之前的 state 等。

（5）useMemo()函数返回一个备忘（memoized）值。对支持备忘的函数来说，调用备忘得到的结果会被保存并缓存起来。以后如果使用相同的输入调用函数，返回的是缓存的值。把创建函数和依赖项数组作为参数传入 useMemo()函数，useMemo()函数仅会在某个依赖项改变时才重新计算备忘值。不要在 useMemo()函数内部执行与渲染无关的操作，如副作用。如果没有提供依赖项数组，useMemo()函数在每次渲染时都会计算新的值。

（6）useCallback()函数返回一个备忘回调函数。把内联回调函数及依赖项数组作为参数传入 useCallback()函数，它将返回该回调函数的备忘版本。该回调函数仅在某个依赖项改变时才会更新。

（7）useRef()函数返回一个可变的 ref 对象，其 current 属性被初始化为传入的参数

（initialValue）。返回的 ref 对象在组件的整个生命周期内保持不变。如果将 ref 对象 MyRef 以<div ref={myRef} /> 形式传入组件属性 ref，无论该节点如何改变，React 都会将 ref 对象的 current 属性设置为相应的 DOM 节点。useRef()函数会在每次渲染时返回同一个 ref 对象。当 ref 对象内容发生变化时，useRef()函数并不会发出通知。如果想要在 React 绑定或解绑 DOM 节点的 ref 对象时运行某些代码，就需要使用回调来实现。

（8）useImperativeHandle()函数使得在使用 ref 对象时可以自定义暴露给父组件的实例值。在大多数情况下，应当避免使用 ref 对象的命令式代码。该函数应当与 forwardRef()方法一起使用。

（9）useLayoutEffect()函数与 useEffect()函数相同，但它会在所有的 DOM 变更之后同步调用 effect。useLayoutEffect()函数在渲染的特定时刻被调用，按照渲染、调用 useLayoutEffect()函数，把组件元素添加到 DOM 中（即浏览器绘制），调用 useEffect()函数的顺序执行。可以使用 useLayoutEffect()函数来读取 DOM 布局并同步触发重渲染。在浏览器执行绘制之前，useLayoutEffect()函数内部的更新计划将被同步刷新。开发时优先用 useEffect()函数，只有当它出现问题的时候再尝试使用 useLayoutEffect()函数。如果使用服务端渲染，无论 useLayoutEffect()函数还是 useEffect()函数都无法在 Javascript 代码加载完成之前执行。

（10）useDebugValue()函数可用于在开发工具中显示自定义 Hook 的标签。React 官方不推荐向每个自定义 Hook 添加 debug 值。当它作为共享库的一部分时才最有价值。在某些情况下，useDebugValue()函数返回格式化的显示值可能是一项开销很大的操作。除非需要检查 Hook，否则没有必要这么做。

8.1.3 自定义 Hook

React 内置了一些像 useState()函数这样的 Hook，也可以创建自定义的 Hook 来复用不同组件之间的状态逻辑。

当需要在两个函数之间共享逻辑时，可以把它提取到第三个函数中。组件和 Hook 都是函数，所以也同样适用这种方式提取组件和 Hook。有时候需要在组件之间复用一些状态逻辑。可以使用高阶组件、render props、自定义 Hook 实现这一目标。

如果函数的名字以 use 开头并调用其他 Hook，就说该函数是一个自定义 Hook。自定义 Hook 必须以 use 开头这个约定非常重要。若不遵循，则会由于无法判断某个函数是否包含对其内部 Hook 的调用，React 将无法自动检查 Hook 是否违反了 Hook 的规则。

可以创建涵盖各种场景的自定义 Hook，如表单处理、动画、订阅声明、计时器等。每次调用 Hook，它都会获取独立的 state。从 React 的角度来看，组件只是调用 useState()函数和 useEffect()函数。在一个组件中多次调用 useState()函数和 useEffect()函数是完全独立的。由于 Hook 本身就是函数，因此可以在它们之间传递信息。

8.1.4 Hook 的使用规则

Hook 是 JavaScript()函数，使用它有两个规则：只能在函数最外层调用 Hook，不要在循环、条件判断或者子函数中调用；只能在 React 的函数组件中调用 Hook，不要在其他

JavaScript 函数中调用。还可以在自定义的 Hook 中调用 Hook。遵循这些规则，确保组件的状态逻辑在代码中清晰可见。

8.2 State Hook 的应用

8.2.1 创建项目 reactjsbook

使用 WebStorm 创建 React 项目 reactjsbook，即选择 React 项目并输入项目地址（采用默认地址）后保持其他内容（如提前装好的 Node.js 和采用默认的脚手架 create-react-app）不变，如图 8-1 所示，再单击图 8-1 所示中的 Create 按钮，等待项目创建完成后，项目基础目录和文件如图 8-2 所示。

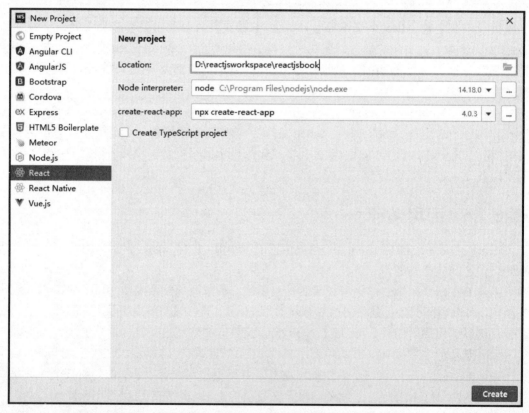

图 8-1　创建 React 项目 reactjsbook 并选择项目信息的界面

对如图 8-2 所示的 public 目录下自动生成的文件 index.html 的代码在删除自动生成的注释并增加一行注释后，如例 8-1 所示。与例 1-3 或例 7-1 所示的代码对照可以看出，此项目中的 index.html 和前面章节中的 index.html 文件作用相同，都是为了提供内容渲染的挂载节点。

【例 8-1】　对文件 index.html 的代码在删除自动生成的注释并增加一行注释后的代码。

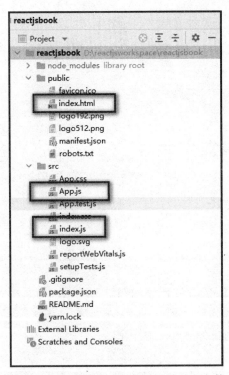

图 8-2 项目创建完成后的基础目录和文件

```
<!DOCTYPE html>
<html lang="en">
  <head>
    <meta charset="utf-8" />
    <link rel="icon" href="%PUBLIC_URL%/favicon.ico" />
    <meta name="viewport" content="width=device-width, initial-scale=1" />
    <meta name="theme-color" content="#000000" />
    <meta
      name="description"
      content="Web site created using create-react-app"
    />
    <link rel="apple-touch-icon" href="%PUBLIC_URL%/logo192.png" />
    <link rel="manifest" href="%PUBLIC_URL%/manifest.json" />
    <title>React App</title>
  </head>
  <body>
    <noscript>You need to enable JavaScript to run this app.</noscript>
<!--下一行代码是整个文件关键代码,请和例 1-3 或例 7-1 所示的代码对照来看-->
    <div id="root"></div>
  </body>
</html>
```

对如图 8-2 所示的 src 目录中自动生成的文件 index.js 的代码在删除注释并增加一行注释后,如例 8-2 所示。与例 1-4 所示的代码或其他章节的 index.js 对照可以看出,此项目中的 index.js 和前面章节中的 index.js 文件作用相同(进行内容的渲染)。不同点在于,此处要渲染的组件 App 是作为一个独立的文件存在。

【例 8-2】 对文件 index.js 的代码在删除自动生成的注释并增加一行注释后的代码。

```
import React from 'react';
import ReactDOM from 'react-dom';
import './index.css';
import App from './App';   //导入组件 App
import reportWebVitals from './reportWebVitals';
//下面 6 行代码是关键代码，用于将组件 App 在 root 节点处渲染出来
ReactDOM.createRoot(document.getElementById('root')).render(
  <React.StrictMode>
    <App />
  </React.StrictMode>
);
reportWebVitals();
```

项目自动生成的 App.js 的文件代码如例 8-3 所示，与之相关的 App.css 主要是对组件 App 的样式控制文件。

【例 8-3】 自动生成的文件 App.js 的代码。

```
import logo from './logo.svg';
import './App.css';
function App() {
  return (
    <div className="App">
      <header className="App-header">
        <img src={logo} className="App-logo" alt="logo" />
        <p>
          Edit <code>src/App.js</code> and save to reload.
        </p>
        <a
          className="App-link"
          href="https://reactjs.org"
          target="_blank"
          rel="noopener noreferrer"
        >
          Learn React
        </a>
      </header>
    </div>
  );
}
export default App;
```

例 8-3 所示中的 export default 表示导出一个任何 JavaScript 类型（如函数），导出的类型可以供其他模块使用。JavaScript 模块是一组可以复用的代码，往往将其保存在单独的文件中，一个文件一个模块。ES 6 中可以用 export default 导出一个类型，也可用 export 导出

多个类型。在一个文件中通过 import 语句导入另一个模块。

如图 8-2 所示的目录 src 中，自动生成的文件 index.css 与控制组件样式的 App.css 的内容一起来控制应用的样式，index.css 和 App.css 的代码运行后的效果，和例 7-2 所示的 index.css 相同。

8.2.2 修改文件 index.js

修改项目 reactjsbook 的 src 目录下自动生成的文件 index.js 的代码如例 8-4 所示。

【例 8-4】 修改后的文件 index.js 的代码。

```
import React from 'react';
import ReactDOM from 'react-dom';
import './index.css';
import App from './App';  //导入组件 App
import reportWebVitals from './reportWebVitals';
import HelloHook from "./components/HelloHook";
ReactDOM.createRoot(document.getElementById('root')).render(
  <React.StrictMode>
    <HelloHook />
  </React.StrictMode>
);
reportWebVitals();
```

8.2.3 创建组件

在项目 reactjsbook 的 src 目录下创建 components 子目录，在 reactjsbook\src\components 目录下创建文件 HookExample1.js，代码如例 8-5 所示。

【例 8-5】 文件 HookExample1.js 的代码。

```
//引入 React 中的 useState Hook，这样就可以在函数组件中存储内部 state
import { useState } from 'react';
const infoMap= {
    beginInfo:'您已经单击了',
    endInfo:'次按钮。',
    btnInfo:'计数器',
}
function HookExample1() {
    // 声明一个新的 state 变量 count
    const [count, setCount] = useState(0);
    return (
        <div>
            <p>{infoMap.beginInfo}{count}{infoMap.endInfo}</p>
            <button onClick={() => setCount(count + 1)}>
                {infoMap.btnInfo}
            </button>
        </div>
    );
```

```
}
export default HookExample1;
```

在 reactjsbook\src\components 目录下创建文件 HelloHook.js，代码如例 8-6 所示。

【例 8-6】 文件 HelloHook.js 的代码。

```
import HookExample1 from "./HookExample1";
function HelloHook() {
    return (
        <div>
            <HookExample1/>
        </div>
    );
}
export default HelloHook;
```

8.2.4 运行项目 reactjsbook

使用 npm start 命令运行项目 reactjsbook 后，控制台的部分输出效果如图 8-3 所示。在浏览器地址栏中输入 localhost:3000，效果如图 8-4 所示。单击图 8-4 中的"计数器"按钮，效果如图 8-5 所示。每单击 1 次"计数器"按钮，"计数器"按钮上面一行的文本中的计数值就加 1。

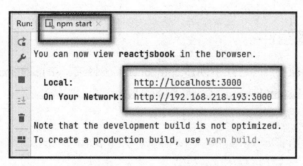

图 8-3 运行项目 reactjsbook 后控制台的部分输出效果

图 8-4 在地址栏中输入 localhost:3000 后浏览器的效果

图 8-5 单击图 8-4 所示中的"计数器"按钮的效果

8.2.5　useState()函数的应用说明

例 8-5 代码中通过调用 useState()函数声明了一个新的 state 变量 count（也可以是其他名字，如 banana）。useState()函数里面唯一的参数就是初始 state。可以按照需要使用数字或字符串对其进行赋值，而不一定是对象。在示例中，只需使用数字来记录用户单击次数，所以传了 0 作为变量的初始 state。如果想在 state 中存储两个不同的变量，只需要调用 useState()函数两次即可。

useState()函数是一种在函数调用时保存变量的方式。一般来说，在函数退出后，变量就会消失，而 state 中的变量会被 React 保留。useState()函数向命名的变量返回一对值。返回值为当前 state（如例 8-5 所示中的 count）以及更新 state 的函数（如例 8-5 所示中的 setCount()函数）。React 会重新渲染组件，并把最新的 state 传给它。这就是写 const [count, setCount] = useState() 的原因。

例 8-5 所示中的语句"const [count, setCount] = useState(0)"的等号左边方括号内的名字并不是 React API 的部分，可以自己取名字；例如"const [fruit, setFruit] = useState('banana');"。这种 JavaScript 语法叫作数组解构。它意味着同时创建了 fruit 和 setFruit 两个值。其中，fruit 的值为 useState()函数返回的第一个值，setFruit 是返回的第二个值。它等价于例 8-7 所示的代码。

【例 8-7】 数组解构的代码。

```
var fruitStateVariable = useState('banana');    // 返回一个有两个元素的数组
var fruit = fruitStateVariable[0];              // 数组里的第一个值
var setFruit = fruitStateVariable[1];           // 数组里的第二个值
```

例 8-7 所示中使用 fruitStateVariable[0]和 fruitStateVariable[1]来访问两个值容易让人产生困惑，因为它们有特定的含义。这就是使用数组解构的原因。

利用数组解构的语法，在调用 useState()函数时，可以给 state 变量取不同的名字。当多次调用 useState()函数时，每次渲染它们的调用顺序是不变的。

8.2.6　State Hook 的等价实现

在 reactjsbook\src\components 目录下创建文件 SameExample1.js，代码如例 8-8 所示。

【例 8-8】 文件 SameExample1.js 的代码。

```
import React from 'react';
const infoMap= {
    beginInfo:'您已经单击了',
    endInfo:'次按钮。',
    btnInfo:'计数器',
}
export default class SameExample extends React.Component {
    constructor(props) {
        super(props);
        this.state = {
            count: 0
```

```
            };
        }
        render() {
            return (
                <div>
                    <p>{infoMap.beginInfo}{this.state.count}{infoMap.endInfo}</p>
                    <button onClick={() => this.setState({ count: this.state.count + 1 })}>
                        {infoMap.btnInfo}
                    </button>
                </div>
            );
        }
    }
```

修改 reactjsbook\src\ 目录下文件 HelloHook.js,代码如例 8-9 所示。

【例 8-9】 修改后的文件 HelloHook.js 的代码。

```
import SameExample1 from "./SameExample1";
function HelloHook() {
    return (
        <div>
            <SameExample1/>
        </div>
    );
}
export default HelloHook;
```

再次执行 npm start 命令运行项目 reactjsbook,运行后的效果如图 8-3 所示。请对照例 8-5 与例 8-8、例 8-6 与例 8-9,分析 Hook 和类组件实现的差异。

视频讲解

8.3　State Hook 的综合应用

8.3.1　创建组件

在 reactjsbook\src\components 目录下创建文件 HookExample11.js,代码如例 8-10 所示。

【例 8-10】 文件 HookExample11.js 的代码。

```
import { useState } from 'react';
const infoMap= {
    beginInfo:'您已经单击了',
    endInfo:'次按钮。',
    btnInfo:'计数器',
    beforeInfo:'猴子已经吃了',
    afterInfo:'根香蕉',
    ageInfo:'年龄',
```

```jsx
    fruitInfo:'水果',
    doInfo:'待办事项:',
}
export default function HookExample11() {
    // 声明一个新的 state 变量 count
    const [count, setCount] = useState(0);
    // 声明多个 state 变量
    const [banana, setBanana] = useState(0);
    const [age, setAge] = useState(42);
    const [fruit, setFruit] = useState('banana');
    const [todos] = useState([{ text: '学习 Hook' }]);
    function handleOrangeClick() {
        // 和 this.setState({ fruit: 'orange' }) 类似
        if(fruit==='banana') {
            setFruit('orange');
        } else {
            setFruit('banana');
        }
    }
    function handleDoClick() {
        alert('学习 React')
    }
    return (
        <div>
            <div>{infoMap.beginInfo}{count}{infoMap.endInfo}</div>
            <button onClick={() => setCount(count + 1)}>
                {infoMap.btnInfo}
            </button>
            <div>{infoMap.beforeInfo}{banana}{infoMap.afterInfo}</div>
            <button onClick={() => setBanana(banana + 1)}>
                {infoMap.btnInfo}
            </button>
            <div>{infoMap.ageInfo}{age}</div>
            <button onClick={() => setAge(age + 1)}>
                {infoMap.ageInfo}
            </button>
            <div>{infoMap.fruitInfo}{fruit}</div>
            <button onClick={handleOrangeClick}>
                {infoMap.fruitInfo}
            </button>
            <p>{infoMap.doInfo}</p>
            <button onClick={handleDoClick}>
                {infoMap.doInfo}
            </button>
        </div>
    );
}
```

在 reactjsbook\src\components 目录下创建文件 HookExample111.js，代码如例 8-11 所示。

【例 8-11】 文件 HookExample111.js 的代码。

```
import { useState } from 'react';
const infoMap= {
    beginInfo:'您已经单击了',
    endInfo:'次按钮。',
    btnInfo:'计数器',
    beforeInfo:'猴子已经吃了',
    afterInfo:'根香蕉',
    ageInfo:'年龄',
    fruitInfo:'水果',
    doInfo:'待办事项',
}
export default function HookExample111() {
    let fruitStateVariable = useState('banana'); // 返回一个包含两个元素的数组
    let fruit = fruitStateVariable[0];           // 数组里的第一个值
    let setFruit = fruitStateVariable[1];        // 数组里的第二个值
    function handleOrangeClick() {
        if(fruit==='banana') {
            setFruit('orange');
        } else {
            setFruit('banana');
        }
    }
    return (
        <div>
            <p>{infoMap.fruitInfo}{fruit}</p>
            <button onClick={handleOrangeClick}>
                {infoMap.fruitInfo}
            </button>
        </div>
    );
}
```

修改 reactjsbook\src\components 目录下文件 HelloHook.js，代码如例 8-12 所示。

【例 8-12】 修改后的文件 HelloHook.js 的代码。

```
import HookExample1 from "./HookExample1";
import SameExample1 from "./SameExample1";
import HookExample11 from "./HookExample11";
import HookExample111 from "./HookExample111";
function HelloHook() {
    return (
        <div>
            <h1>State Hook 的综合示例</h1>
            <hr/>
            <HookExample1/>
            <hr/>
            <h1>等价实现计数器</h1>
```

```
            <SameExample1/>
            <hr/>
            <h1>多个 state 值的应用开发</h1>
            <HookExample11/>
            <hr/>
            <h1>state Hook 的数组解构实现</h1>
            <HookExample111/>
        </div>
    );
}
export default HelloHook;
```

8.3.2　运行项目 reactjsbook

使用 npm start 命令运行项目 reactjsbook，在浏览器地址栏中输入 localhost:3000，效果如图 8-6 所示。单击图 8-6 所示中的各个按钮，执行效果请读者在自己操作的基础上或在参考视频中进行观察。

图 8-6　运行项目 reactjsbook 后在地址栏中输入 localhost:3000 的效果

8.4　Effect Hook 的应用

8.4.1　说明

在 React 组件中，函数在正常返回之外做的执行数据获取、订阅或者手动修改 DOM

等操作被称为副作用(effect)。在 React 组件中有需要清除和不需要清除两类常见的副作用。发送网络请求、手动变更 DOM、记录日志等都是常见的无须清除的操作；还有一些副作用是需要清除的，例如，订阅外部数据源，在这种情况下，清除工作是非常重要的，可以防止引起内存泄漏。

useEffect()函数能给函数组件增加操作副作用的能力，如等待渲染结束把值提供给 alert()方法和 console.log()方法等。它跟类组件中的 componentDidMount()方法、componentDidUpdate()方法和 componentWillUnmount()方法具有相同的功能，useEffect()函数是三种方法合并成的一个 API。

调用 useEffect()函数表明在完成对 DOM 的更改后运行新增的副作用函数。因为副作用函数是在组件内声明的，所以它们可以访问到组件的 props 和 state。在默认情况下，React 会在每次渲染（包括第一次）后调用副作用函数。

可以使用多个 useEffect()函数将不相关逻辑分离到不同的 effect 中。React 将按照 effect 声明的顺序依次调用组件中的每个 effect。如果某些特定值在两次重渲染之间没有发生变化，就可以让 React 跳过对 effect()函数的调用。

8.4.2　创建文件 HookExample2.js

在 reactjsbook\src\components 目录下创建文件 HookExample2.js，代码如例 8-13 所示。

【例 8-13】 文件 HookExample2.js 的代码。

```
import {useEffect, useState} from 'react';
const infoMap= {
    beginInfo: '您已经单击了',
    endInfo: '次按钮。',
    btnInfo: '计数器',
}
function HookExample2() {
    const [count, setCount] = useState(0);
    // 相当于 componentDidMount()方法和 componentDidUpdate()方法
    useEffect(() => {
    // 使用浏览器的 API 更新页面标题
        document.title = infoMap.beginInfo+`${count}`+infoMap.endInfo;
    });
    return (
        <div>
            <div>{infoMap.beginInfo}{count}{infoMap.endInfo}</div>
            <button onClick={() => setCount(count + 1)}>
                {infoMap.btnInfo}
            </button>
        </div>
    );
}
export default HookExample2;
```

例 8-13 所示的代码是对例 8-5 所示的代码的修改，为计数器增加了一个小功能：将

document 的 title 设置为包含了点击次数的消息。

例 8-13 所示代码中的 useEffect()函数表明需要在渲染后执行某些操作。React 会保存传递的函数，并且在执行 DOM 更新之后调用它。将 useEffect()函数放在组件内部，可以在 effect 中直接访问 state 变量 count（或其他 props）。它已经保存在函数作用域中，不需要特殊的 API 来读取它。

在每次渲染中传递给 useEffect()函数都会有所不同。事实上，这正是可以在 effect 中获取最新的 count 的值，而不用担心其过期的原因。每次重新渲染，都会生成新的 effect 以替换之前的 effect。从某种意义上讲，effect 更像是渲染结果的一部分，每个 effect 都属于一次特定的渲染。

与 componentDidMount()方法或 componentDidUpdate()方法不同，使用 useEffect()函数调度的 effect 不会阻塞浏览器更新 UI 界面，这让应用看起来响应更快。在大多数情况下，effect 不需要同步地执行。在个别情况下（例如测量布局），有 Hook 的 useLayoutEffect()函数可供使用，其 API 与 useEffect()函数相同。

8.4.3 Effect Hook 的等价实现

在 reactjsbook\src\components 目录下创建文件 SameExample2.js，代码如例 8-14 所示。

【例 8-14】 文件 SameExample2.js 的代码。

```
import React from 'react';
const infoMap= {
    beginInfo:'您已经单击了',
    endInfo:'次按钮。',
    btnInfo:'计数器',
}
export default class sameExample2 extends React.Component {
    constructor(props) {
        super(props);
        this.state = {
            count: 0
        };
    }
    componentDidMount() {
        document.title = infoMap.beginInfo+'${this.state.count}'+infoMap.endInfo;
    }
    componentDidUpdate() {
        document.title = infoMap.beginInfo+'${this.state.count}'+infoMap.endInfo;
    }
    render() {
        return (
            <div>
                <p>{infoMap.beginInfo}{this.state.count}{infoMap.endInfo}</p>
```

```
            <button onClick={()=>this.setState({count:this.state.count + 1 })}>
                {infoMap.btnInfo}
            </button>
        </div>
    );
    }
}
```

在类组件中，render()方法不应该有任何副作用。一般来说，希望在 React 更新 DOM 之后才执行自定义的操作。这就是把副作用操作放到 componentDidMount()方法和 componentDidUpdate()方法中的原因。

8.4.4 创建组件

在 reactjsbook\src\components 目录下创建文件 FriendStatusHook.js，代码如例 8-15 所示。

【例 8-15】 文件 FriendStatusHook.js 的代码。

```
import { useState, useEffect } from 'react';
const infoMap= {
    idInfo:1,
    subscribeToFriendStatusInfo:'Hook 中的 subscribeToFriendStatus 方法',
    unsubscribeFromFriendStatusInfo:'Hook 中的 unsubscribeFromFriendStatus 方法',
    loadInfo:'装载...',
    onInfo:'在线',
    offInfo:'离线'
}
function subscribeToFriendStatus(id, handleStatusChange,status){
    console.log(infoMap.idInfo);
    console.log(infoMap.subscribeToFriendStatusInfo);
    handleStatusChange(status);
}
function unsubscribeFromFriendStatus(id, handleStatusChange,status){
    console.log(infoMap.idInfo);
    console.log(infoMap.unsubscribeFromFriendStatusInfo);
    handleStatusChange(status);
}
export default function FriendStatusHook(props) {
    const [isOnline, setIsOnline] = useState(null);
    useEffect(() => {
        function handleStatusChange(status) {
            setIsOnline(status.isOnline);
        }
        // eslint-disable-next-line no-restricted-globals
        subscribeToFriendStatus(props.id, handleStatusChange,status);
        return function cleanup() {
            // eslint-disable-next-line no-restricted-globals
```

```
            unsubscribeFromFriendStatus(props.id,handleStatusChange, status);
        };
    });
    if (isOnline === null) {
        return infoMap.loadInfo;
    }
    return isOnline ? infoMap.onInfo : infoMap.offInfo;
}
```

useEffect()函数的设计是在同一个地方执行清除操作。如果 effect 返回一个函数，React将会在执行清除操作时调用它，并在组件卸载的时候执行清除操作。这是 effect 可选的清除机制。每个 effect 都可以返回一个清除函数。

在 reactjsbook\src\components 目录下创建文件 SameFriendStatus.js，代码如例 8-16 所示。

【例 8-16】 文件 SameFriendStatus.js 的代码。

```
import React from 'react';
const infoMap= {
    idInfo:1,
    subscribeToFriendStatusInfo:'class 中的 subscribeToFriendStatus 方法',
    unsubscribeFromFriendStatusInfo:'class 中的 unsubscribeFromFriendStatus 方法',
    loadInfo:'装载...',
    onInfo:'在线',
    offInfo:'离线'
}
function subscribeToFriendStatus(id, handleStatusChange,status){
    console.log(infoMap.idInfo);
    console.log(infoMap.subscribeToFriendStatusInfo);
}
function unsubscribeFromFriendStatus(id, handleStatusChange,status){
    console.log(infoMap.idInfo);
    console.log(infoMap.unsubscribeFromFriendStatusInfo);
}
export default class SameFriendStatus extends React.Component {
    constructor(props) {
        super(props);
        this.state = { isOnline: null };
        this.handleStatusChange = this.handleStatusChange.bind(this);
    }
    componentDidMount() {
        subscribeToFriendStatus(this.props.id, this.handleStatusChange, this.status);
    }
    componentWillUnmount() {
        unsubscribeFromFriendStatus(this.props.id, this.handleStatusChange, this.status);
    }
```

```
        handleStatusChange(status) {
            this.setState({
                isOnline: status.isOnline
            });
        }
        render() {
            if (this.state.isOnline === null) {
                return infoMap.loadInfo;
            }
            return this.state.isOnline ? infoMap.onInfo : infoMap.offInfo;
        }
    }
```

注意，componentDidMountt()方法和 componentWillUnmountt()方法之间是相互对应的。使用生命周期方法要求拆分这些逻辑代码，即使这两部分代码都有相同的副作用。例 8-16 所示的代码还需要编写 componentDidUpdate t()方法才能保证完全正确。

在 reactjsbook\src\components 目录下创建文件 HelloHook2.js，代码如例 8-17 所示。

【例 8-17】 文件 HelloHook2.js 的代码。

```
import HookExample2 from "./HookExample2";
import SameExample2 from "./SameExample2";
import FriendStatusHook from "./FriendStatusHook";
import SameFriendStatus from "./SameFriendStatus";
export default function HelloHook2() {
    return (
        <div>
            <h1>Effect Hook 的综合示例</h1>
            <hr/>
            <HookExample2/>
            <hr/>
            <h1>等价实现计数器</h1>
            <SameExample2/>
            <hr/>
            <h1>需要清除的 effect</h1><hr/>
            <FriendStatusHook id={1}/>
            <h1>需要清除的 effect 的等价实现</h1>
            <SameFriendStatus id={1}/>
        </div>
    );
}
```

8.4.5 修改文件 index.js

修改 reactjsbook\src 目录下文件 index.js 的代码，如例 8-18 所示。

【例 8-18】 修改后的文件 index.js 的代码。

```
import React from 'react';
import ReactDOM from 'react-dom';
```

```
import './index.css';
import reportWebVitals from './reportWebVitals';
import HelloHook2 from "./components/HelloHook2";
ReactDOM.createRoot(document.getElementById('root')).render(
  <React.StrictMode>
    <HelloHook2/>
  </React.StrictMode>
);
reportWebVitals();
```

8.4.6 运行项目 reactjsbook

使用 npm start 命令运行项目 reactjsbook，在浏览器中输入 localhost:3000，打开浏览器的"开发者工具"功能，效果如图 8-7 所示。图 8-7 的细节和单击图 8-7 所示中的各个按钮的效果请读者在自己操作的基础上或在参考视频中进行观察。

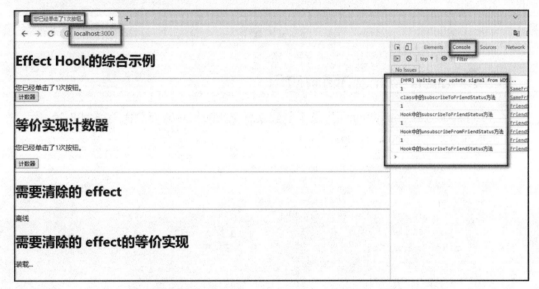

图 8-7　运行项目后在浏览器中输入 localhost:3000 且打开浏览器的"开发者工具"功能的效果

8.5 其他 Hook 的应用

8.5.1 useState()函数应用

在 reactjsbook\src\components 目录下创建文件 CounterHook.js，代码如例 8-19 所示。

【例 8-19】 文件 CounterHook.js 的代码。

```
import {useState} from "react";
const infoMap={
    inParamsInfo:'传入的初始值是：',
    nowInfo:'现在的结果：',
    middleInfo:'，',
```

```
    resetInfo:'重置',
    addInfo:'减一',
    subInfo:'加一',
}
export default function Counter( props) {
    let initialCount=props.initialCount;
    const [count, setCount] = useState(initialCount);
    return (
        <>
            {infoMap.inParamsInfo}{initialCount}{infoMap.middleInfo}
            {infoMap.nowInfo}{count}{infoMap.middleInfo}
            <button onClick={()=>setCount(initialCount)}>{infoMap.resetInfo}</button>
    <button onClick={() => setCount(prevCount => prevCount - 1)}>{infoMap.subInfo}</button>
    <button onClick={() => setCount(prevCount => prevCount + 1)}>{infoMap.addInfo}</button>
        </>
    );
}
```

8.5.2 useReducer()函数应用

在 reactjsbook\src\components 目录下创建文件 ReducerHook.js，代码如例 8-20 所示。

【例 8-20】 文件 ReducerHook.js 的代码。

```
import React, { useReducer } from "react";
const initialState = 0;
const myContext = React.createContext();
function reducer(state, action) {
    switch (action.type) {
        case "reset":
            return initialState;
        case "increment":
            return { count: state.count + 1 };
        case "decrement":
            return { count: state.count - 1 };
        default:
            return state;
    }
}
const ContextProvider = props => {
    const [state, dispatch] = useReducer(reducer, { count: 0 });
    return (
        <myContext.Provider value={{ state, dispatch }}>
            {props.children}
        </myContext.Provider>
    );
};
export { reducer, myContext, ContextProvider };
```

8.5.3　useMemo()函数应用

在 reactjsbook\src\components 目录下创建文件 MemoHook.js，代码如例 8-21 所示。

【例 8-21】　文件 MemoHook.js 的代码。

```
import React, { useState, useMemo } from 'react';
const Child = ({ age, name, children }) => {
    console.log(age, name, children, '11111111');
    function namechange() {
        console.log(age, name, children, '22222222');
        return name + 'change';
    }
    const changedname = useMemo(() => namechange(), [ name ]);
    return (
        <div style={{ border: '1px solid' }}>
            <p>children: {children}</p>
            <p>name: {name}</p>
            <p>changed: {changedname}</p>
            <p>age: {age}</p>
        </div>
    );
};
const MemoHook = () => {
    const [ name, setname ] = useState('baby张');
    const [ age, setage ] = useState(18);
    return (
        <div>
            <button
                onClick={() => {
                    setname('baby张' + new Date().getTime());
                }}
            >
                改名字
            </button>
            <button
                onClick={() => {
                    setage('年龄' + new Date().getTime());
                }}
            >
                改年龄
            </button>
            <p>
                UseMemo {name}: {age}
            </p>
            <Child age={age} name={name}>
                {name}的 children
            </Child>
        </div>
    );
};
```

```
export default MemoHook;
```

8.5.4　useRef()函数应用

在 reactjsbook\src\components 目录下创建文件 RefHook.js，代码如例 8-22 所示。

【例 8-22】 文件 RefHook.js 的代码。

```
import React, { useState, useRef } from 'react';
const RefHook = () => {
    const [ name, setname ] = useState('baby张');
    const refvalue = useRef(null);
    function addRef() {
        refvalue.current.value = name;
        console.log(refvalue.current.value);
    }
    return (
        <div>
            <input
                defaultValue={name}
                onChange={(e) => {
                    setname(e.target.value);
                }}
            />
            <button onClick={addRef}>给下面插入名字</button>
            <p>给我个 UseRef 名字：</p>
            <input ref={refvalue} />
        </div>
    );
};
export default RefHook;
```

8.5.5　创建组件

在 reactjsbook\src\目录下创建文件 HelloHook3.js，代码如例 8-23 所示。

【例 8-23】 文件 HelloHook3.js 的代码。

```
import CounterHook from "./CounterHook";
import {ContextProvider} from "./ReducerHook";
import MemoHook from "./MemoHook";
import RefHook from "./RefHook";
export default function HelloHook3() {
    return (
        <div>
            <h3>Hook API 的综合示例</h3>
            <CounterHook initialCount={1}/>
            <h3>useReducer</h3>
            <ContextProvider children={2}/>
            <h3>useMemo</h3>
            <MemoHook />
            <h3>useRef</h3>
```

```
        <RefHook/>
      </div>
  );
}
```

8.5.6 修改文件 index.js

修改 reactjsbook\src 目录下文件 index.js 的代码，如例 8-24 所示。

【例 8-24】 修改后的文件 index.js 的代码。

```
import React from 'react';
import ReactDOM from 'react-dom';
import './index.css';
import reportWebVitals from './reportWebVitals';
import HelloHook3 from "./components/HelloHook3";
ReactDOM.createRoot(document.getElementById('root')).render(
  <React.StrictMode>
    <HelloHook3/>
  </React.StrictMode>
);
reportWebVitals();
```

8.5.7 运行项目 reactjsbook

使用 npm start 命令运行项目 reactjsbook，在浏览器中输入 localhost:3000，效果如图 8-8

图 8-8　运行项目后在浏览器中输入 localhost:3000 的效果

所示。在图 8-8 所示的文本框中输入内容和单击图 8-8 所示中的各个按钮的效果请读者在自己操作的基础上或在参考视频中进行观察。

习题 8

一、简答题

1. 简述对 Hook 的理解。
2. 简述对常见 Hook API 的理解。
3. 简述对常见自定义 Hook 的理解。
4. 简述对 Hook 使用规则的理解。

二、实验题

1. 完成 State Hook 的应用开发。
2. 完成 Effect Hook 的应用开发。
3. 完成其他 Hook API 的应用开发。

第 9 章

React基础原理和高级指引

本章先简要介绍 React 基础原理（使用 React、JSX 表示对象、类组件的执行顺序、异步编程、Fiber、模块）、React 应用开发的一般步骤，再介绍 React 片段（fragment）、context、高阶组件、ref 转发、portal、ref 和 DOM、Web Component、render props、错误边界、测试等内容。

9.1 React 基础原理

视频讲解

9.1.1 选择性地使用 React

React 从一开始就被设计为逐步采用，开发人员可以根据需要选择性地使用 React。若想在已有应用程序的页面中局部地添加交互性，使用 React 组件是一种不错的选择。通过仅仅几行代码并且无需使用构建工具，就可以在应用程序的一小部分中使用 React。然后，再逐步扩展它的使用范围。或者，只将其涵盖在少数动态部件中。

通常来说，每个新的 React 应用程序的顶层组件都是 App 组件。但是，如果将 React 集成到已有的应用程序中，可能需要使用像 Button 这样的小组件，并自下而上地将这类组件逐步应用到更大范围。

9.1.2 JSX 表示对象

Babel 会把 JSX 转译成对 React.createElement()方法的调用。例 9-1 和例 9-2 所示的代码功能完全等效。

【例 9-1】 创建 element1 的代码。

```
const element1 = (
  <h1 className="greeting">
```

```
    Hello, world!
  </h1>
);
```

【例 9-2】 创建 element2 的等效代码。

```
const element2= React.createElement(
  'h1',
  {className: 'greeting'},
  'Hello, world!'
);
```

React.createElement()方法会预先执行一些检查,以减少代码中的错误,它会创建一个如例 9-3 中 element3 所示的对象。

【例 9-3】 创建 element3 的等效代码。

```
// 注意：这是简化过的结构
const element 3 = {
  type: 'h1',
  props: {
    className: 'greeting',
    children: 'Hello, world!'
  }
};
```

三段等效代码的综合示例如例 9-4 所示。先在项目 firstreact 根目录下创建子目录 insidereactjs,在 firstreact\insidereactjs 目录下创建子目录 jsxequalelement,在 firstreact\insidereactjs\jsxequalelement 目录下创建文件 jsxequalelement.html,代码与例 1-3 所示的代码相同。在 firstreact\insidereactjs\jsxequalelement 目录下创建文件 index.js,代码如例 9-4 所示。

【例 9-4】 在 firstreact\insidereactjs\jsxequalelement 目录下创建的文件 index.js 的代码。

```
const divReact = document.getElementById('root');
const infoMap= {
    typeInfo:'h1',
    classSelectName:'greeting',
    helloInfo:'Hello JSX 等价示例',
}
const  element1= (
    <infoMap.typeInfo className={infoMap.classSelectName}>
        {infoMap.helloInfo}
    </infoMap.typeInfo>
);
const element2= React.createElement(
    infoMap.typeInfo,
    {className: infoMap.classSelectName},
    infoMap.helloInfo
);
const element3 = {
    type: infoMap.typeInfo,
```

```
        props: {
            className: infoMap.classSelectName,
            children: infoMap.helloInfo
        }
    };
    function ElementComp1(){
        return element1;
    }
    function ElementComp2(){
        return element2;
    }
    function ElementComp3(){
        return <element3.type>{element3.props.children}</element3.type>;
    }
    const example = (
    <div>
        <ElementComp1/>
        <ElementComp2/>
        <ElementComp3/>
    </div>
    );
    ReactDOM.render(example, divReact);
```

运行文件 jsxequalelement.html，效果如图 9-1 所示。

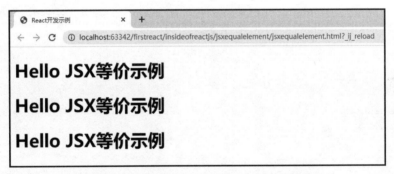

图 9-1　运行文件 jsxequalelement.html 的效果

9.1.3　类组件的执行顺序

在 firstreact\insidereactjs 目录下创建子目录 componentsteps，在 firstreact\insidereactjs\componentsteps 目录下创建文件 componentsteps.html，代码与例 1-3 所示的代码相同。在 firstreact\insidereactjs\componentsteps 目录下创建文件 index.js，代码如例 9-5 所示。

【例 9-5】 在 firstreact\insidereactjs\componentsteps 目录下创建的文件 index.js 的代码。

```
const divReact = document.getElementById('root');
const infoMap= {
    helloInfo:'组件的步骤',
    timeInfo:'现在时间是：',
    endInfo:'。'
```

```jsx
    }
    class Clock extends React.Component {
        constructor(props) {
            super(props);
            this.state = {date: new Date()};
        }
        componentDidMount() {
            this.timerID = setInterval(
                () => this.tick(),
                1000
            );
        }
        componentWillUnmount() {
            clearInterval(this.timerID);
        }
        tick() {
            this.setState({
                date: new Date()
            });
        }
        render() {
            return (
                <div>
                    <h1>{infoMap.helloInfo}</h1>
                    <h2>{infoMap.timeInfo}{this.state.date.toLocaleTimeString()}{infoMap.endInfo}</h2>
                </div>
            );
        }
    }
    const stepComp = (
        <Clock/>
    );
    ReactDOM.render(stepComp, divReact);
```

运行文件 componentsteps.html，效果如图 9-2 所示。注意，图 9-2 所示中的时间是运行文件时的时间。

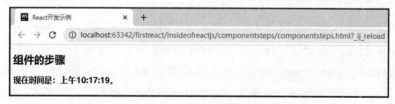

图 9-2　运行文件 componentsteps.html 的效果

组件的执行顺序（以类组件为例）包括如下五个步骤。

（1）当组件（如 Clock 组件）被传给 root.render()方法时，React 会调用组件的构造方法。因为 Clock 组件需要显示当前的时间，所以它用一个包含当前时间的对象 date 来初始化 this.state。

（2）React 调用组件（如 Clock 组件）的 render()方法来渲染组件。确定在组件上应该展示什么内容；然后 React 通过更新 DOM 渲染出组件的输出效果。

（3）当组件（如 Clock 组件）的输出被插入 DOM 中后，React 就会调用 componentDidMount()方法。在这个方法中，Clock 组件向浏览器请求设置一个计时器来每秒调用一次组件的 tick()方法。

（4）在浏览器调用的 tick()方法中，Clock 组件会通过调用 setState()方法来计划进行一次 UI 界面更新。setState()方法的调用让 React 知道 state 发生了变化，React 会重新调用 render()方法来确定页面上该显示什么。此时 render()方法中的 this.state.date 会重新渲染输出更新过的时间。React 也会相应的更新 DOM。

（5）一旦 Clock 组件从 DOM 中被移除，React 就会调用 componentWillUnmount()方法，并停止计时器。

9.1.4　异步编程

在 Web 应用中通常存在异步任务。在 JavaScript 中，异步任务不可以影响主线程，在等待异步任务完成的时间段内，JavaScript 可以完成其他任务。

promise 是异步任务的一种解决方案。promise 可以获取异步操作的消息，即异步操作处在挂起（或称为初始、未定）、完成或失败等某种状态。从语法上讲，promise 是一个代理对象，被代理的值在 promise 对象在创建时可能是未知的。使用 fetch()方法可以处理简单的 promise。但注意，fetch()方法必须至少接收一个参数（所请求资源的路径），其中包括请求方式、请求头、请求体等参数。

处理 promise 的另一种方法是创建异步函数。异步函数要用关键字 async 声明，在成功处理 promise 之前暂停执行异步函数中的后续代码，promise 在调用前需要加上关键字 await，可让异步函数等待 promise 被成功处理。

异步函数的结果可能成功或出现错误。在 fetch()方法的基础上可以使用.then()方法作为成功处理完 promise 后的回调函数（即调用的方法），可以串接多个.then()方法处理成功的 promise，可以使用.catch()方法作为处理 promise 失败后的回调函数。在使用 async 和 await 时，要把 promise 调用放在 try…catch 块中以处理可能出现的错误。也可以串接.then()方法作为成功处理 promise 后的回调函数，如果 promise 被拒绝，相关细节可以发回给.catch()方法或 try…catch 块中。

9.1.5　Fiber

由于更新 DOM 是同步操作，要等到所有更新完成并渲染出来之后，主线程才能开始执行其他任务，此更新操作是一项耗时的任务，会影响到用户的体验。为了解决这个问题，React 在 16.0 版中使用了 Fiber，采用异步方式重写了 DOM 的更新机制。React 16.0 版分离了渲染器和调和器（reconciler）。其中，调和器（调和算法）保留在 React Core 包中；渲染器则负责具体渲染，React DOM、React Native 等负责处理渲染逻辑，可以接入 React 的调和算法。

Fiber 把较长时间阻塞的更新操作（称为 work）分成较小的单元（fiber）。一个 fiber 是一个 JavaScript 对象，负责记录正在调和什么及其在更新循环中的什么位置。

当一个 fiber 结束后，React 检查主线程中有没有什么重要的事情要做。如果有重要的工作，React 就交出主线程的控制权。当重要的工作做完之后，React 继续更新。如果主线程接下来没有重要的任务，React 就转到下一个 fiber，然后在 DOM 中渲染改动。

9.1.6 模块

JavaScript 模块（module）是一组可以重用的代码，易于插入 JavaScript 文件而不产生变量冲突。JavaScript 模块保存在独立的文件中，一个文件一个模块。

ES 6 开始可以用 export default 表示导出一个任何 Javascript 类型，如原始类型、对象、数组和函数，也可以用 export 导出多个类型。导出的类型可以供其他模块使用。在一个文件中通过 import 语句导入另一个模块。

ES 6 中的 import 和 export 还没有得到所有浏览器和 Node.js 的完全支持，而通过 Babel 可以实现完全支持。除此之外，也可以用 CommonJS 模式中的 module.exports 来描述模块的导出和使用 require 描述模块的导入。CommonJS 是所有版本 Node.js 都支持的模块模式，Babel 和 webpack 也支持此种模块。

在 ES 6 出现之前，模块加载方案主要有 CommonJS 和 AMD（Asynchronous Module Definition，异步模块定义）两种。其中，CommonJS 主要用于服务器，AMD 主要用于浏览器。AMD 采用异步加载的方式加载模块，避免模块加载阻塞网页的渲染进度。AMD 作为一个规范，只需定义其语法 API，而不关心其实现。AMD 是 RequireJS 在推广过程中对模块定义的规范化。RequireJS 从 2.0 开始可以延迟执行。RequireJS 的模块语法允许它尽快地加载多个模块，虽然加载的顺序不定，但依赖的顺序最终是正确的。同时因为无须创建全局变量，甚至可以做到在同一个页面上同时加载同一模块的不同版本。

9.2 React 应用开发的一般步骤

9.2.1 将 UI 界面分解为组件

在设计稿上对设计好的 UI 界面用方框圈出每一个组件（包括它们的子组件），并且以合适的名称命名组件。

可以根据单一功能原则（或称为单一职责原则）来分解组件（提取组件）。一个组件原则上只能负责一个功能。因为应用程序经常是在向用户展示 JSON 数据模型，如果数据模型设计得恰当，UI 界面（及其对应的组件树）便会与数据模型一一对应，UI 界面和数据模型都会倾向于遵守相同的信息结构。将 UI 界面分解为组件，每个组件会与数据模型的某部分匹配。

9.2.2 实现应用程序的静态版本

若确定了组件，则可以实现应用程序。将渲染 UI 界面和添加交互两个过程分开，即先用已有的数据模型渲染一个不包含交互功能的 UI 界面（即应用程序的静态版本）再添加交互功能。在实现静态版本时往往要编写大量代码，而不需要考虑太多交互细节；在添加交互功能时，则要考虑大量细节，而不需要编写太多代码。

在实现静态版本时，要创建一些复用其他组件的组件，然后通过 props 传入所需的数据。注意，不要使用 state 来实现静态版本。因为 state 代表随时间会产生变化的数据，所以其一般在添加交互功能时使用。

可以用自上而下或者自下而上的方法实现应用程序。自上而下意味着先实现层级较高的组件，自下而上意味着先实现最基础的小组件。对于比较简单的应用程序，使用自上而下的方法更方便；对于较为大型的应用程序，自下而上的方法更加简单。

至此，组件目前只需提供 render() 方法用于渲染。顶层的组件通过 props 接收数据模型。如果数据模型发生了改变，再次调用 render() 方法，UI 界面就会相应地被更新。通过数据模型变化—调用 render() 方法—UI 界面发生相应变化的过程很容易看清 UI 界面在哪里被更新以及如何被更新的。

9.2.3　确定 state

想要使 UI 界面具备交互功能，需要有触发改变基础数据模型的能力。React 通过 state 来完成这个任务。

实现应用程序时，需要找出应用所需的可变 state 的最小集合，并根据需要计算出其他所有数据。判断依据是 DRY（Don't Repeat Yourself，不要重复你自己）原则。对于由父组件通过 props 传递而来的数据、随着时间的推移而保持不变的数据、能够根据其他 state 或 props 计算出的数据等不是 state。除此之外的数据可以设定为 state。

9.2.4　确定 state 的放置位置

确定了所需的 state 的最小集合之后，需要确定哪个组件能够改变这些 state，或者说拥有这些 state。

React 中的数据是顺着组件层级从上往下传递的。在找到根据某个 state 进行渲染的所有组件的基础上，找出在组件层级上高于所有需要该 state 的组件的最近父组件（简称最近父组件）。拥有 state 所有组件的最近父组件或者比它（最近父组件）层级更高的组件可以存放 state（即状态提升）。如果找不到一个合适的位置来存放该 state，可以直接创建一个新的组件来存放，并将这一新组件置于最近父组件的位置。

9.2.5　添加反向数据流

至此，借助于自上而下传递的 props 和 state 渲染了一个应用程序。接着，可以尝试让数据反向传递，即通过处于较低层级的表单组件更新较高层级的组件中的 state。

由于 state 只能由拥有它们的组件进行更改，每当需要改变子组件 state 时，父组件必须将一个能够触发 state 改变的回调函数传递给子组件。可以使用事件来监视子组件 state 的变化，并通知父组件传递给子组件的回调函数；然后该回调函数将调用 setState() 方法，从而更新应用。

9.3 React 片段

9.3.1 说明

React 中不能在一个组件中渲染两个及以上（多个）毗邻元素或同辈元素时，必须将多个元素放在一个标签内（如<div> </div>）。这样会导致创建大量非必需的标签，产生很多没有什么真正用途的容器。此时使用 React 片段（fragment）可以解决此问题，可以模拟容器的行为，但是不真正创建新的标签，无须向 DOM 添加额外节点。

React 片段允许使用一种新的更简短的语法来(<> </>)声明片段。它看起来像空标签。它除了不支持 key 或属性，可以像任何其他元素一样使用；也可以显式使用<React.Fragment>语法声明的片段，此时片段可以拥有 key。此种语法的一个使用场景是将一个集合映射到一个片段数组。key 是唯一可以传递给片段的属性。

9.3.2 创建组件

在 reactjsbook\src\components 目录下创建文件 FragmentsExample.js，代码如例 9-6 所示。

【例 9-6】 文件 FragmentsExample.js 的代码。

```
import React from "react";
class Table extends React.Component {
    render() {
        return (
            <table>
                <tr>
                    <Columns />
                </tr>
            </table>
        );
    }
}
class Columns extends React.Component {
    render() {
        return (
            <div>
                <td>Hello</td>
                <td>World</td>
            </div>
        );
    }
}
class Columns2 extends React.Component {
    render() {
```

```
            return (
                <React.Fragment>
                    <td>Hello</td>
                    <td>World</td>
                </React.Fragment>
            );
        }
    }
    class Columns3 extends React.Component {
        render() {
            return (
                <>
                    <td>Hello</td>
                    <td>World</td>
                </>
            );
        }
    }
    function Glossary(props) {
        return (
            <dl>
                {props.items.map(item => (
                    // 如果没有 key，那么 React 会发出一个关键警告
                    <React.Fragment key={item.id}>
                        <dt>{item.term}</dt>
                        <dd>{item.description}</dd>
                    </React.Fragment>
                ))}
            </dl>
        );
    }
    export { Table,Columns3,Columns2,Columns,Glossary} ;
```

例 9-6 所示的代码中，<Columns />需要返回多个<td>元素直接插入<tr>和</tr>中以使渲染的 HTML 有效。如果如例 9-6 所示在<Columns />的 render()方法中使用了父标签<div>，那么生成的 HTML 达不到预期的效果。

在 reactjsbook\src\components 目录下创建文件 HelloFragments.js，代码如例 9-7 所示。

【例 9-7】 文件 HelloFragments.js 的代码。

```
import {Columns, Columns2, Columns3, Glossary, Table} from "./FragmentsExample";
import React from "react";
const items = [{
    id: 1,
    term: 'term1',
    description: 'desc1'
},
{
```

```
            id: 2,
            term: 'term2',
            description: 'desc2'
        },
]
export default function HelloFragments() {
    return (
        <div>
            <h3>Fragments 示例</h3>
            <Table/>
            <h3>Columns</h3>
            <Columns/>
            <h3>Columns2</h3>
            <Columns2/>
            <h3>Columns3</h3>
            <Columns3/>
            <h3>Glossary</h3>
            <Glossary items={items}/>
        </div>
    );
}
```

9.3.3 修改文件 index.js

修改 reactjsbook\src 目录下文件 index.js 的代码，如例 9-8 所示。

【例 9-8】 修改后的文件 index.js 的代码。

```
import React from 'react';
import ReactDOM from 'react-dom';
import './index.css';
import reportWebVitals from './reportWebVitals';
import HelloFragments from "./components/HelloFragments";
ReactDOM.createRoot(document.getElementById('root')).render(
  <React.StrictMode>
    <HelloFragments/>
  </React.StrictMode>
);
reportWebVitals();
```

9.3.4 运行项目 reactjsbook

使用 npm start 命令运行项目 reactjsbook，在浏览器中输入 localhost:3000，效果如图 9-3 所示。

图 9-3　运行项目后在浏览器中输入 localhost:3000 的效果

9.4　context

9.4.1　说明

React 中数据通常是通过 props 由父组件向子组件传递的。这种做法对于传递某些类型数据过于烦琐。例如，地区偏好、首选语言、UI 界面主题、当前认证的用户或者一些缓存数据，这些数据是应用程序中许多组件都需要的相同数据。context（上下文）提供了一种在组件之间共享此类全局数据的方法，而不必显式地通过组件树的逐层传递 props。context 能让这些数据向组件树下所有的组件进行广播，所有组件都能访问到这些数据，也能访问到后续的数据更新。

如果只是想避免逐层传递一些属性（数据），有时组件的组合与 context 相比，是更好的解决方案。对组件的控制反转减少了要传递的 props 数量，这在很多场景下会使得代码更加干净，对根组件有更多的把控。将逻辑提升到组件树的更高层次来处理的方法会使得这些高层组件变得更复杂，而且组件并不限制于接收单个子组件。可能会传递多个子组件，甚至会为这些子组件封装多个单独的接口（slot）。

为了把数据放入 context，就要创建 context 的提供者（provider），context 的提供者可以把数据（对象）放入 context 中。但是它自己无法修改上下文中的值，需要父级组件的协助。从上下文中获取数据的组件称为消费者（consumer）。 React 渲染一个订阅了 context 的组件时，这个组件会从组件树中离自身最近的那个匹配的提供者中读取 context 值。当组件所处的树中没有匹配到提供者时，其 defaultValue 参数会生效。将 undefined 传递给提供者的 value 时，消费者组件的 defaultValue 不会生效。

每个 context 对象都会返回一个提供者——React 组件，它允许消费者组件订阅 context 的变化。提供者接收一个 value 属性传递给消费者。一个提供者可以和多个消费者组件有对应关系。多个提供者也可以嵌套使用，里层的会覆盖外层的数据。当提供者 value 值发生变化时，它的所有消费者都会重新渲染。提供者及其消费者都不受制于 shouldComponentUpdate()方法，因此当消费者在其祖先组件退出更新的情况下也能更新。

挂载在类上的 contextType 属性会被赋值为一个由 React.createContext() 方法创建的 context 对象。这使得能使用 this.context 来消费最近 context 的值。可以在任何生命周期方法中访问到它，包括在 render()方法中。如果两个或者更多的 context 值经常被一起使用，就要考虑创建自定义的渲染组件，以提供这些值。因为 context 会使用参考标识来决定何时进行渲染，这里可能会有一些陷阱，当提供者的父组件进行重新渲染时，可能会在消费者中触发意外的渲染。

9.4.2 创建组件

在 reactjsbook\src\components 目录下创建文件 ContextExample.js，代码如例 9-9 所示。

【例 9-9】 文件 ContextExample.js 的代码。

```
import React from "react";
import {Button} from "react-bootstrap";   //安装 react-bootstrap
class AppContext1 extends React.Component {
    render() {
        return <Toolbar1 theme="dark" />;
    }
}
function Toolbar1(props) {
    // Toolbar1 组件接收一个额外的 theme(主题)属性，然后传递给 ThemedButton 组件
    // 如果应用中每一个单独的按钮都需要知道 theme 的值，这会是件很麻烦的事
    // 因为必须将这个值层层传递给所有组件
    return (
        <div>
            <ThemedButton1 theme={props.theme} />
        </div>
    );
}
class ThemedButton1 extends React.Component {
    render() {
        return <Button theme={this.props.theme} >Context1</Button>;
    }
}
// 利用 context，可以无须明确地传遍每一个组件，就能将值传递进组件树
// 为当前的 theme 创建一个 context，默认值为 light
const ThemeContext = React.createContext('light');
class AppContext2 extends React.Component {
    render() {
```

```
            // 使用一个 Provider 来将当前的 theme 传递给组件树
            // 无论组件树的层次有多深，任何组件都能读取这个值
            // 将 dark 作为当前的值传递下去
            return (
                <ThemeContext.Provider value="dark">
                    <Toolbar2 />
                </ThemeContext.Provider>
            );
    }
}
// 中间的组件再也不必指明往下传递 theme 了
function Toolbar2() {
    return (
        <div>
            <ThemedButton2 />
        </div>
    );
}
const themes = {
    light: {
        foreground: '#000000',
        background: '#eeeeee',
    },
    dark: {
        foreground: '#cc0033',
        background: '#222222',
    },
};
const ThemeContext2 = React.createContext(
    themes.dark // 默认值
);
class ThemedButton2 extends React.Component {
    // 指定 contextType 读取当前的 theme context
    // React 会往上找到最近的 theme Provider，然后使用它的值
    //当前的 theme 值为 dark
    static contextType = ThemeContext2;
    render() {
        return <Button theme={this.context}>Context2</Button>;
    }
}
export { AppContext1,AppContext2 } ;
```

在 reactjsbook\src\components 目录下创建文件 HelloContext.js，代码如例 9-10 所示。

【例 9-10】 文件 HelloContext.js 的代码。

```
import React from "react";
import {AppContext1, AppContext2} from "./ContextExample";
```

```
export default function HelloContext() {
    return (
        <div>
            <h3>Context 示例</h3>
            <h3>AppContext1</h3>
            <AppContext1/>
            <h3>AppContext2</h3>
            <AppContext2/>
        </div>
    );
}
```

9.4.3 修改文件 index.js

修改 reactjsbook\src 目录下文件 index.js 的代码,如例 9-11 所示。

【例 9-11】 修改后的文件 index.js 的代码。

```
import React from 'react';
import ReactDOM from 'react-dom';
import './index.css';
import reportWebVitals from './reportWebVitals';
import HelloContext from "./components/HelloContext";
ReactDOM.createRoot(document.getElementById('root')).render(
  <React.StrictMode>
      <HelloContext/>
  </React.StrictMode>
);
reportWebVitals();
```

9.4.4 运行项目 reactjsbook

使用 npm start 命令运行项目 reactjsbook,在浏览器中输入 localhost:3000,效果如图 9-4 所示。

图 9-4 运行项目后在浏览器中输入 localhost:3000 的效果

9.5 高阶组件

9.5.1 说明

高阶组件（HOC）是 React 中复用组件逻辑的一种高级技巧。高阶组件不是 React API 的一部分，它是一种基于 React 的组合特性而形成的设计模式。高阶组件是参数为组件（函数）、返回值为组件（函数）或两者兼备的组件（函数）。

一般的组件是将 props 转换为 UI 界面，而高阶组件是将组件转换为另一个组件。HOC 不会修改传入的组件，也不会使用继承来复制其行为。HOC 通过将组件包装在另一个容器组件中来组成新组件。HOC 没有副作用。被包装组件接收来自容器组件的所有 props，同时也接收组件自身的用于渲染的 props。HOC 不关心数据的使用方式或原因，被包装组件也不关心数据是怎么来的。

容器组件担任分离高层逻辑和低层逻辑的责任，由容器组件管理订阅和状态，并将 props 传递给处理渲染的 UI 界面。HOC 返回的组件与原组件应保持类似的接口。

React 的调和算法使用组件标识来确定它是应该更新已有子树还是将其丢弃并挂载新的子树。如果从 render()方法返回的组件与前一个渲染中的组件相同（===），那么 React 根据旧子树与新子树的不同来递归更新子树。如果它们不相等，就直接卸载前一个子树。因此，不要在组件的 render()方法中对一个组件应用 HOC。这不仅仅会有性能问题，重新挂载该组件时还会导致该组件及其所有子组件的状态丢失。

如果在组件之外创建 HOC，组件只会被创建一次。因此，在每次渲染时都会是同一个组件。在极少数情况下，需要动态调用 HOC。可以在组件的生命周期方法或其构造方法中进行调用。

虽然高阶组件的约定是将所有 props 传递给被包装组件，但这对于 ref 并不适用。因为 ref 实际上并不是一个 props，它是由 React 专门处理的对象。如果将 ref 添加到 HOC 的返回组件中，那么 ref 引用指向容器组件，而不是被包装组件。

9.5.2 创建组件

在 reactjsbook\src\components 目录下创建文件 HOCExample.js，代码如例 9-12 所示。

【例 9-12】 文件 HOCExample.js 的代码。

```
import React from "react";
class Base1 extends React.Component {
    render() {
        return "hello,text!"
    }
}
// HOC 函数实现 2.0 版
const toUpperCaseHoc = function(WrappedComponent) {
    return class Hoc extends React.Component {
```

```
        render() {
            const { text } = this.props;
            const text2Upper = text.toUpperCase();
            return <WrappedComponent text={text2Upper} />;
        }
    };
};
// 实现 1.0 版
class Base2 extends React.Component {
    render() {
        return this.props.text;
    }
}
// 用 HOC 包装后生成的新的组件，符合 2.0 版需求，同时包含了 1.0 版的其他功能
const HelloWorld2Upper = toUpperCaseHoc(Base2);
export {Base1,Base2,HelloWorld2Upper}
```

在 reactjsbook\src\components 目录下创建文件 HelloHOC.js，代码如例 9-13 所示。

【例 9-13】 文件 HelloHOC.js 的代码。

```
import React from "react";
import {Base1, HelloWorld2Upper} from "./HOCExample";
export default function HelloHOC() {
    return (
        <div>
            <hr/>
            <h3>HOC 示例</h3>
            <h3>非高阶组件</h3>
            <Base1/>
            <h3>高阶组件</h3>
            <HelloWorld2Upper text="hello,HOC!" />
        </div>
    )
}
```

9.5.3 修改文件 index.js

修改 reactjsbook\src 目录下文件 index.js 的代码，如例 9-14 所示。

【例 9-14】 修改后的文件 index.js 后的代码。

```
import React from 'react';
import ReactDOM from 'react-dom';
import './index.css';
import reportWebVitals from './reportWebVitals';
import HelloContext from "./components/HelloContext";
import HelloHOC from "./components/HelloHOC";
ReactDOM.createRoot(document.getElementById('root')).render(
    <React.StrictMode>
        <HelloContext/>
```

```
        <HelloHOC/>
    </React.StrictMode>
);
reportWebVitals();
```

9.5.4 运行项目 reactjsbook

使用 npm start 命令运行项目 reactjsbook，在浏览器中输入 localhost:3000，效果如图 9-5 所示。

图 9-5 运行项目后在浏览器中输入 localhost:3000 的效果

9.6 ref 转发

9.6.1 说明

ref 转发是一个可选特性，是一项将 ref 自动地通过组件传递到其子组件的技巧，其允许某些组件使用 React.forwardRef() 来获取、接收传递给它的 ref，并将其向下传递（转发它）给子组件。

React 组件隐藏其实现细节，包括其渲染结果。其他使用该组件的组件通常不需要获取内部的 DOM 元素的 ref。这可以很好地防止组件过度依赖其他组件的 DOM 结构。虽然这种封装对高层组件比较理想，但对高可复用的叶子组件（底层组件）来说可能不太方便。底层组件倾向于在整个应用中以一种类似常规 DOM 中 button（或 input）等的方式被使用。为了管理焦点、选择文本或制作动画，这些组件必须要访问其 DOM 节点。

对于大多数应用中的组件来说，通常不需要 ref 转发。但其对某些组件，尤其是可复用的组件库是很有用的。

9.6.2 创建组件

在 reactjsbook\src\components 目录下创建文件 RefsExample.js，代码如例 9-15 所示。

【例 9-15】 文件 RefsExample.js 的代码。

```
import React from "react";
function handleClick(){
    alert("Refs 转发示例")
}
//通过调用 React.createRef()方法创建了一个 ref，并将其赋值给变量 FancyButton
//再通过指定 ref 为 JSX 属性将其向下传递给<button>
const FancyButton = React.forwardRef((props, ref) => (
    <button ref={ref} className="FancyButton" onClick={handleClick}>
        {props.children}
    </button>
));
const Input = InputComponent => {
    const forwardRef = (props, ref) => {
        const onType = () => console.log(ref.current.value);
        return<InputComponent forwardedRef={ref}onChange={onType}{...props}/>;
    };
    return React.forwardRef(forwardRef);
};
const TextInput = ({ forwardedRef, children, ...rest }) => (
    <div>
        <input ref={forwardedRef} {...rest} />
        {children}
    </div>
);
const InputField = Input(TextInput);
class CustomTextInput extends React.Component {
    render() {
        const inputRef = React.createRef();
        return <InputField ref={inputRef} />;
    }
}
export {FancyButton,CustomTextInput}
```

在 reactjsbook\src\components 目录下创建文件 HelloRefs.js，代码如例 9-16 所示。

【例 9-16】 文件 HelloRefs.js 的代码。

```
import React from "react";
import {CustomTextInput, FancyButton} from "./RefsExample";
const ref = React.createRef();
export default function HelloRefs() {
    return (
        <div>
```

```
        <hr/>
        <h3>Refs 转发示例</h3>
        <FancyButton ref={ref} >Refs 转发示例</FancyButton>
        <h3>高阶组件 Refs 转发示例</h3>
        <CustomTextInput />
        </div>
    )
}
```

9.6.3 修改文件 index.js

修改 reactjsbook\src 目录下文件 index.js 的代码,如例 9-17 所示。

【**例 9-17**】 修改后的文件 index.js 的代码。

```
import React from 'react';
import ReactDOM from 'react-dom';
import './index.css';
import reportWebVitals from './reportWebVitals';
import HelloContext from "./components/HelloContext";
import HelloHOC from "./components/HelloHOC";
import HelloRefs from "./components/HelloRefs";
ReactDOM.createRoot(document.getElementById
('root')).render(
    <React.StrictMode>
        <HelloContext/>
        <HelloHOC/>
        <HelloRefs/>
    </React.StrictMode>
);
reportWebVitals();
```

图 9-6 运行项目后在浏览器中输入 localhost:3000 的效果

9.6.4 运行项目 reactjsbook

使用 npm start 命令运行项目 reactjsbook,在浏览器中输入 localhost:3000,效果如图 9-6 所示。

9.7 portal

9.7.1 说明

通常,当从组件的 render()方法返回一个元素时,该元素将被挂载到 DOM 节点中离其最近的父节点。然而有时希望将子元素插入 DOM 节点中的不同位置。例如,当父组件有 overflow: hidden 或 z-index 样式时需要子组件能够在视觉上跳出其容器(如对话框、悬浮卡、提示框等)。portal 提供了一种将子节点渲染到存在于父组件以外的 DOM 节点的

视频讲解

方法。使用 portal 时，管理键盘焦点就变得尤为重要。

除了可以被放置在 DOM 树中任何地方的特性之外，portal 和普通的 React 子节点行为一致。由于 portal 仍存在于 DOM 树中，且与 DOM 树中的位置无关，所以无论其子节点是否是 portal，其 context、事件冒泡等功能特性都是不变地在父组件里捕获一个来自 portal 冒泡上来的事件，使之能够在开发时具有不完全依赖于 portal 的更为灵活的抽象。

9.7.2 创建组件

在 reactjsbook\src\components 目录下创建文件 PortalsExample.js，代码如例 9-18 所示。

【例 9-18】 文件 PortalsExample.js 的代码。

```javascript
import React from "react";
import * as ReactDOM from "react-dom";
const modalRoot = document.getElementById('root');
class Modal extends React.Component {
    constructor(props) {
        super(props);
        this.el = document.createElement('div');
    }
    componentDidMount() {
        // Modal 所有子元素被挂载后，portal 元素会被嵌入 DOM 树中
        // 子元素将被挂载到一个分离的 DOM 节点中
        // 如果要求子组件在挂载时可以立刻接入 DOM 树
        // 例如衡量一个 DOM 节点，或者在后代节点中使用 autoFocus
        // 就需将 state 添加到 Modal 中，仅当 Modal 被插入 DOM 树中才能渲染子元素
        modalRoot.appendChild(this.el);
    }
    componentWillUnmount() {
        modalRoot.removeChild(this.el);
    }
    render() {
        return ReactDOM.createPortal(
            this.props.children,
            this.el
        );
    }
}
class Parent extends React.Component {
    constructor(props) {
        super(props);
        this.state = {clicks: 0};
        this.handleClick = this.handleClick.bind(this);
    }
    handleClick() {
        // 当子元素里的按钮被单击时，这个将会被触发更新父元素的 state
        // 即使这个按钮在 DOM 中不是直接关联的后代
```

```
            this.setState(state => ({
                clicks: state.clicks + 1
            }));
        }
        render() {
            return (
                <div onClick={this.handleClick}>
                    <p>单击按钮的次数：{this.state.clicks}</p>
                    <p>
                        打开浏览器的"开发者工具"功能观察 button 的归属
                    </p>
                    <Modal>
                        <Child />
                    </Modal>
                </div>
            );
        }
    }
    function Child() {
        // 这个按钮的单击事件会冒泡到父元素，因为这里没有定义 onClick 属性
        return (
            <div className="modal">
                <button>Click</button>
            </div>
        );
    }
    export {Parent}
```

在 reactjsbook\src\components 目录下创建文件 HelloPortals.js，代码如例 9-19 所示。

【例 9-19】 文件 HelloPortals.js 的代码。

```
import React from "react";
import {Parent} from "./PortalsExample";
export default function HelloPortals() {
    return (
        <div>
            <hr/>
            <h3>Portals 示例</h3>
            <Parent/>
        </div>
    )
}
```

9.7.3 修改文件 index.js

修改 reactjsbook\src 目录下文件 index.js 的代码如例 9-20 所示。

【例 9-20】 修改后的文件 index.js 的代码。

```
import React from 'react';
import ReactDOM from 'react-dom';
import './index.css';
import reportWebVitals from './reportWebVitals';
import HelloContext from "./components/
HelloContext";
import HelloHOC from "./components/HelloHOC";
import HelloRefs from "./components/HelloRefs";
import HelloPortals from "./components/
HelloPortals";
ReactDOM.createRoot(document.getElementById
('root')).render(
    <React.StrictMode>
        <HelloContext/>
        <HelloHOC/>
        <HelloRefs/>
        <HelloPortals/>
    </React.StrictMode>
);
reportWebVitals();
```

9.7.4 运行项目 reactjsbook

使用 npm start 命令运行项目 reactjsbook，在浏览器中输入 localhost:3000，效果如图 9-7 所示。

图 9-7 运行项目后在浏览器中输入 localhost:3000 的效果

9.8 ref 和 DOM

9.8.1 说明

ref 是使用 React.createRef()方法创建的属性（对象），并通过 ref 属性附加到 React 元素。在构造组件时，通常将 ref 分配给实例属性，以便可以在整个组件中引用它们。

适合使用 ref 的场景包括管理焦点、文本选择或媒体播放、触发强制动画、集成第三方 DOM 库等情况。要避免使用 ref 来做任何可以通过声明式编程完成的事情。例如，在 Dialog 组件中传递 isOpen 属性而不要暴露 open()方法。

当 ref 属性用于 HTML 元素时，构造方法中使用 ref 接收底层 DOM 元素作为其 current 属性。当 ref 属性用于自定义类组件时，ref 对象接收组件的挂载实例作为其 current 属性。不能在函数组件上使用 ref 属性，因为它们没有实例。

在 React 16.3 及其更高版本中，ref 转发使得组件可以像暴露自己的 ref 一样暴露子组件的 ref。React 也支持回调 ref。它能更精细地控制何时设置和解除 ref。不同于 ref 属性，回调 ref 会传递一个函数。这个函数中接收 React 组件实例或 HTML DOM 元素作为参数，以使它们能在其他地方被存储和访问。在组件挂载时，会调用 ref 回调函数并传入 DOM 元素，当卸载时调用它并传入 null。在 componentDidMount()方法或 componentDidUpdate()方法触发前，React 会保证 ref 一定是最新的。可以在组件间传递回调形式的 ref，就像可以传递 ref 一样。如果 ref 回调函数是以内联函数的方式定义的，在更新过程中它会被执行两次。其中，第一次传入参数 null，第二次传入 DOM 元素（参数）。

9.8.2 创建组件

在 reactjsbook\src\components 目录下创建文件 RefsAndDOMEx.js，代码如例 9-21 所示。

【例 9-21】 文件 RefsAndDOMEx.js 的代码。

```
import React from "react";
//创建 ref
class MyComponent extends React.Component {
    constructor(props) {
        super(props);
        this.myRef = React.createRef();
    }
    render() {
        return <div ref={this.myRef} />;
    }
}
//为 DOM 元素添加 ref
//在组件挂载时给 current 属性传入 DOM 元素，并在组件卸载时传入 null 值
//ref 会在 componentDidMount()方法 或 componentDidUpdate()方法触发前更新
class CustomTextInput extends React.Component {
    constructor(props) {
        super(props);
        // 创建一个 ref 来存储 textInput 的 DOM 元素
        this.textInput = React.createRef();
        this.focusTextInput = this.focusTextInput.bind(this);
    }
    focusTextInput() {
        // 直接使用原生 API 使 text 输入框获得焦点，通过 current 来访问 DOM 节点
        this.textInput.current.focus();
    }
    render() {
        // 通知 React 把 <input> ref 关联到构造方法里创建的 textInput 上
        return (
            <div>
                <input
                    type="text"
                    ref={this.textInput} />
                <input
                    type="button"
                    value="前面的文本框获得焦点"
                    onClick={this.focusTextInput}
                />
```

```jsx
            </div>
        );
    }
}
//为类组件添加 ref
class AutoFocusTextInput extends React.Component {
    constructor(props) {
        super(props);
        this.textInput = React.createRef();
    }
    componentDidMount() {
        this.textInput.current.focusTextInput();
    }
    render() {
        return (
            <CustomTextInput ref={this.textInput} />
        );
    }
}
//回调 ref
class CustomTextInput2 extends React.Component {
    constructor(props) {
        super(props);
        this.textInput = null;
        this.setTextInputRef = element => {
            this.textInput = element;
        };
        this.focusTextInput = () => {
            // 使用原生 DOM API 使 text 输入框获得焦点
            if (this.textInput) this.textInput.focus();
        };
    }
    componentDidMount() {
        // 组件挂载后,让文本框自动获得焦点
        this.focusTextInput();
    }
    render() {
        // 使用 ref 的回调函数将 text 输入框 DOM 节点的引用存储到 React
        return (
            <div>
                <input
                    type="text"
                    ref={this.setTextInputRef}
                />
                <input
                    type="button"
                    value="前面的文本框获得焦点"
                    onClick={this.focusTextInput}
                />
```

```
            </div>
        );
    }
}
function CustomTextInput3(props) {
    return (
        <div>
            <input ref={props.inputRef} />
        </div>
    );
}
class Parent extends React.Component {
    render() {
        return (
            <CustomTextInput3
                inputRef={el => this.inputElement = el}
            />
        );
    }
}
export {MyComponent,AutoFocusTextInput,CustomTextInput2,Parent}
```

在 reactjsbook\src\components 目录下创建文件 HelloRefsAndDOM.js，代码如例 9-22 所示。

【例 9-22】文件 HelloRefsAndDOM.js 的代码。

```
import React from "react";
import{AutoFocusTextInput,CustomTextInput2,MyComponent}from "./RefsAndDOMEx";
import {Parent} from "./PortalsExample";
export default function HelloRefsAndDOM() {
    return (
        <div>
            <hr/>
            <h3>Refs 和 DOM 示例</h3>
            <h3>示例 1</h3>
            <MyComponent/>
            <h3>示例 2</h3>
            <AutoFocusTextInput/>
            <h3>示例 3</h3>
            <CustomTextInput2/>
            <h3>示例 4</h3>
            <Parent/>
        </div>
    )
}
```

9.8.3 修改文件 index.js

修改 reactjsbook\src 目录下文件 index.js 的代码如例 9-23 所示。

【例 9-23】 修改后的文件 index.js 的代码。

```
import React from 'react';
import ReactDOM from 'react-dom';
import './index.css';
import reportWebVitals from './reportWebVitals';
import HelloRefsAndDOM from "./components/HelloRefsAndDOM";
ReactDOM.createRoot(document.getElementById('root')).render(
  <React.StrictMode>
     <HelloRefsAndDOM/>
  </React.StrictMode>
);
reportWebVitals();
```

9.8.4 运行项目 reactjsbook

使用 npm start 命令运行项目 reactjsbook,在浏览器中输入 localhost:3000,效果如图 9-8 所示。

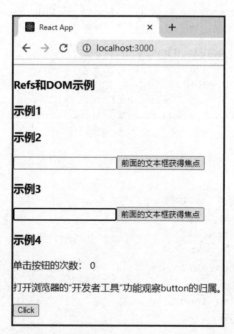

图 9-8 运行项目后在浏览器中输入 localhost:3000 的效果

9.9 Web Component

9.9.1 说明

React 和 Web Component 是为了解决不同的问题而产生的。Web Component 为可复用组件提供了强大的封装。React 则提供了声明式编程的解决方案，使 DOM 与数据保持同步。两者旨在互补。可以在 Web Component 中使用 React，或者在 React 中使用 Web Component。通常在 React 的应用开发中不需要使用 Web Component，只有在使用 Web Component 编写的第三方 UI 界面组件时，可能需要用到 Web Component。

Web Component 通常暴露的是命令式 API。例如，Web Component 的组件 video 可能会公开 play() 和 pause() 方法。要访问 Web Component 的命令式 API，需要使用 ref 直接与 DOM 节点进行交互。如果使用的是第三方 Web Component，最好的解决方案是编写 React 组件来包装该 Web Component。Web Component 触发的事件可能无法通过 React 渲染树正确的传递。需要在 React 组件中手动添加事件处理器来处理这些事件。

9.9.2 创建组件

在 reactjsbook\src\components 目录下创建文件 WebComponentsEx.js，代码如例 9-24 所示。

【例 9-24】 文件 WebComponentsEx.js 的代码。

```
import React from "react";
import * as ReactDOM from "react-dom";
class HelloMessage extends React.Component {
    render() {
        return <div>Hello <x-search>{this.props.name}</x-search>!</div>;
    }
}
function BrickFlipbox() {
    return (
        <brick-flipbox class="demo">
            <div>front</div>
            <div>back</div>
        </brick-flipbox>
    );
}
class XSearch extends HTMLElement {
    connectedCallback() {
        const mountPoint = document.createElement('span');
        this.attachShadow({ mode: 'open' }).appendChild(mountPoint);
        const name = this.getAttribute('name');
        const url = 'https://www.google.com/search?q=' + encodeURIComponent(name);
```

```
                ReactDOM.render(<a href={url}>{name}</a>, mountPoint);
        }
}
customElements.define('x-search', XSearch);
export {HelloMessage,BrickFlipbox,XSearch}
```

在 reactjsbook\src\components 目录下创建文件 HelloWebComponents.js，代码如例 9-25 所示。

【例 9-25】 文件 HelloWebComponents.js 的代码。

```
import React from "react";
import {BrickFlipbox, HelloMessage} from "./WebComponentsEx";
export default function HelloWebComponents() {
    return (
        <div>
            <hr/>
            <h3>Web Components 示例</h3>
            <h3>示例 1</h3>
            <HelloMessage name={"React"}/>
            <h3>示例 2</h3>
            <BrickFlipbox/>
        </div>
    )
}
```

9.9.3　修改文件 index.js

修改 reactjsbook\src 目录下文件 index.js 的代码，如例 9-26 所示。

【例 9-26】 修改后的文件 index.js 的代码。

```
import React from 'react';
import ReactDOM from 'react-dom';
import './index.css';
import reportWebVitals from './reportWebVitals';
import HelloWebComponents from "./components/HelloWebComponents";
ReactDOM.createRoot(document.getElementById('root')).render(
  <React.StrictMode>
      <HelloWebComponents/>
  </React.StrictMode>
);
reportWebVitals();
```

9.9.4　运行项目 reactjsbook

使用 npm start 命令运行项目 reactjsbook，在浏览器中输入 localhost:3000，效果如图 9-9 所示。

图 9-9　运行项目后在浏览器中输入 localhost:3000 的效果

9.10　render props

9.10.1　说明

render props 是指在 React 组件之间使用值为函数的 props 从而实现共享代码的简单技术。它是一个用于告知组件需要渲染什么内容的函数。任何被用于告知组件需要渲染什么内容的函数 props 都可以被称为 render props。

具有 render props 的组件可以用于接收一个函数。该函数能够实现返回一个 React 元素并调用它，而不是实现组件的渲染逻辑。提供一个带有函数 props 的组件，它能够动态地决定什么是需要渲染的，并有效地改变它的渲染结果。这项技术使共享行为变得非常容易。要获得这个行为，只要渲染一个带有 render props 的组件就可以了。

9.10.2　创建组件

在 reactjsbook\src\components 目录下创建文件 RenderPropEx.js，代码如例 9-27 所示。准备一个图片文件 Cat.jpg 并将其放在项目的 public 目录下。

【例 9-27】　文件 RenderPropEx.js 的代码。

```
import React from "react";
class MouseTracker1 extends React.Component {
    constructor(props) {
        super(props);
        this.handleMouseMove = this.handleMouseMove.bind(this);
        this.state = { x: 0, y: 0 };
    }
    handleMouseMove(event) {
        this.setState({
            x: event.clientX,
            y: event.clientY
        });
    }
```

```
        render() {
            return (
                <div style={{height: '100vh'}}onMouseMove={this.handleMouseMove}>
                    <h1>移动鼠标!</h1>
                    <p>当前的鼠标位置是 ({this.state.x}, {this.state.y})</p>
                </div>
            );
        }
    }
    // <Mouse> 组件封装了需要的行为
    class Mouse extends React.Component {
        constructor(props) {
            super(props);
            this.handleMouseMove = this.handleMouseMove.bind(this);
            this.state = { x: 0, y: 0 };
        }
        handleMouseMove(event) {
            this.setState({
                x: event.clientX,
                y: event.clientY
            });
        }
        render() {
            return (
                <div style={{height:'100vh'}}onMouseMove={this.handleMouseMove}>
                    {/*渲染 <p> 以外的东西 */}
                    <p>The current mouse position is ({this.state.x}, {this.state.y})</p>
                </div>
            );
        }
    }
    class MouseTracker2 extends React.Component {
        render() {
            return (
                <>
                    <h1>移动鼠标!</h1>
                    <Mouse/>
                </>
            );
        }
    }
    class Cat extends React.Component {
        render() {
            const mouse = this.props.mouse;
            return (
    <img src="../cat.jpg" style={{ position: 'absolute', left: mouse.x, top: mouse.y }} alt="es-lint want to get"/>
            );
```

```
        }
    }
    class MouseWithCat extends React.Component {
        constructor(props) {
            super(props);
            this.handleMouseMove = this.handleMouseMove.bind(this);
            this.state = { x: 0, y: 0 };
        }
        handleMouseMove(event) {
            this.setState({
                x: event.clientX,
                y: event.clientY
            });
        }
        render() {
            return (
                <div style={{ height: '100vh' }} onMouseMove={this.handleMouseMove}>
                    <Cat mouse={this.state}/>
                </div>
            );
        }
    }
    class MouseTracker3 extends React.Component {
        render() {
            return (
                <div>
                    <h1>移动鼠标!</h1>
                    <MouseWithCat/>
                </div>
            );
        }
    }
    class Cat2 extends React.Component {
        render() {
            const mouse = this.props.mouse;
            return (
    <img src="../cat.jpg" style={{position:'absolute',left:mouse.x,top: mouse.y }} alt="es-lint want to get"/>
            );
        }
    }
    class Mouse2 extends React.Component {
        constructor(props) {
            super(props);
            this.handleMouseMove = this.handleMouseMove.bind(this);
            this.state = { x: 0, y: 0 };
        }
        handleMouseMove(event) {
```

```
            this.setState({
                x: event.clientX,
                y: event.clientY
            });
        }
        render() {
            return (
                <div style={{height:'100vh'}}onMouseMove={this.handleMouseMove}>
                    {/*
                    不是提供静态的<Mouse>渲染方法
                     使用render props动态决定渲染的内容
                    */}
                    {this.props.render(this.state)}
                </div>
            );
        }
    }
    class MouseTracker4 extends React.Component {
        render() {
            return (
                <div>
                    <h1>移动鼠标!</h1>
                    <Mouse render={mouse => (
                        <Cat mouse={mouse} />
                    )}/>
                </div>
            );
        }
    }
    // 如果出于某种原因需要HOC,那么可以使用具有render props的普通组件来创建HOC
    function withMouse(Component) {
        return class extends React.Component {
            render() {
                return (
                    <Mouse render={mouse => (
                        <Component {...this.props} mouse={mouse} />
                    )}/>
                );
            }
        }
    }
    class MouseTracker5 extends React.Component {
        render() {
            return (
                <div>
                    <h1>Move the mouse around!</h1>

                    {/*
                    这是不好的做法
```

```
                每个渲染的 render props 的值将会不同
                */}
                <Mouse2 render={mouse => (
                    <Cat2 mouse={mouse} />
                )}/>
            </div>
        );
    }
}
class MouseTracker6 extends React.Component {
    // 定义实例方法.renderTheCat()方法,在渲染中使用它时,它指的是相同的函数
    renderTheCat(mouse) {
        return <Cat mouse={mouse} />;
    }
    render() {
        return (
            <div>
                <h1>Move the mouse around!</h1>
                <Mouse render={this.renderTheCat} />
            </div>
        );
    }
}
export {MouseTracker1,MouseTracker2,MouseTracker3,MouseTracker4,
MouseTracker5,MouseTracker6,withMouse}
```

在 reactjsbook\src\components 目录下创建文件 HelloRenderProp.js,代码如例 9-28 所示。

【例 9-28】 文件 HelloRenderProp.js 的代码。

```
import React from "react";
import {MouseTracker1, MouseTracker2, MouseTracker3, MouseTracker4,
MouseTracker5, MouseTracker6} from "./RenderPropEx";
export default function HelloRenderProp() {
    return (
        <div>
            <hr/>
            <h3>Render Prop 示例</h3>
            <h3>示例 1</h3>
            <MouseTracker1/>
            <h3>示例 2</h3>
            <MouseTracker2/>
            <h3>示例 3</h3>
            <MouseTracker3/>
            <h3>示例 4</h3>
            <MouseTracker4/>
            <h3>示例 5</h3>
            <MouseTracker5/>
            <h3>示例 6</h3>
```

```
            <MouseTracker6/>
        </div>
    )
}
```

9.10.3 修改文件 index.js

修改 reactjsbook\src 目录下文件 index.js 的代码，如例 9-29 所示。

【例 9-29】 修改后的文件 index.js 的代码。

```
import React from 'react';
import ReactDOM from 'react-dom';
import './index.css';
import reportWebVitals from './reportWebVitals';
import HelloRenderProp from "./components/HelloRenderProp";
ReactDOM.createRoot(document.getElementById('root')).render(
  <React.StrictMode>
      <HelloRenderProp/>
  </React.StrictMode>
);
reportWebVitals();
```

9.10.4 运行项目 reactjsbook

使用 npm start 命令运行项目 reactjsbook，在浏览器中输入 localhost:3000，部分效果如图 9-10 所示。

图 9-10　运行项目后在浏览器中输入 localhost:3000 的部分效果

9.11　错误边界

9.11.1　说明

在 React 16.0 版本之前，组件内的 JavaScript 错误会导致 React 的内部状态被破坏，并且在下一次渲染时可能产生无法追踪的错误。这些错误基本上是由应用程序中已有的其他代码（非 React 组件代码）引起的，但是 React 并没有提供一种在组件中优雅处理这些错误的方法，也无法从错误中恢复。项目越复杂，组件树越庞大，代码就越难以调试。为了解决这个问题，在 React 16.0 版本中引入了错误边界（Error Boundaries）这个概念。

错误边界是一种 React 组件，这类组件可以捕获并打印发生在其子组件树任何位置的 JavaScript 错误，并且它会渲染出备用 UI 界面，而不是渲染那些崩溃了的子组件树。错误边界在渲染期间、生命周期方法和整个组件树的构造方法中捕获错误。错误边界无法捕获事件处理、异步编程、服务端渲染、组件自身（并非组件的子组件）抛出来的错误。

如果一个类组件中定义了 getDerivedStateFromError()方法或 componentDidCatch()方法中的任意一个（或两个）时，那么它就变成一个错误边界。当抛出错误后，可以用 getDerivedStateFromError()方法渲染备用 UI 界面，用 componentDidCatch()方法打印错误信息。

9.11.2　创建组件

在 reactjsbook\src\components 目录下创建文件 ErrorBoundariesEx.js，代码如例 9-30 所示。

【例 9-30】 文件 ErrorBoundariesEx.js 的代码。

```
import React from "react";
class ErrorBoundary extends React.Component {
    constructor(props) {
        super(props);
        this.state = {hasError: false};
    }
    static getDerivedStateFromError(error) {
        return {hasError: true};
    }
    componentDidCatch(error, errorInfo) {
        //也可以将错误日志上报给服务器
        alert(error+errorInfo)
    }
    render() {
        if (this.state.hasError) {
            return <h1>Something went wrong.</h1>;
        }
        return <h1>{this.props.children}</h1>;
    }
}
class MyComponent extends React.Component {
    constructor(props) {
        super(props);
        this.state = { error: null };
        this.handleClick = this.handleClick.bind(this);
    }
    handleClick() {
```

```
            try {
                // 执行操作，若有错误，则会被抛出
            } catch (error) {
                this.setState({ error });
            }
        }
        render() {
            if (this.state.error) {
                return <h1>Caught an error.</h1>
            }
            return <button onClick={this.handleClick}>Click Me</button>;
        }
    }
    export {ErrorBoundary,MyComponent}
```

在 reactjsbook\src\components 目录下创建文件 HelloErrorBoundaries.js，代码如例 9-31 所示。

【例 9-31】 文件 HelloErrorBoundaries.js 的代码。

```
import React from "react";
import {ErrorBoundary, MyComponent} from "./ErrorBoundariesEx";
export default function HelloErrorBoundaries() {
    return (
        <div>
            <h3>Error Boundaries 示例</h3>
            <h3>示例 1</h3>
            <MyComponent/>
            <ErrorBoundary><div>示例 2</div></ErrorBoundary>
        </div>
    );
}
```

9.11.3 修改文件 index.js

修改 reactjsbook\src 目录下文件 index.js 的代码，如例 9-32 所示。

【例 9-32】 修改后的文件 index.js 的代码。

```
import React from 'react';
import ReactDOM from 'react-dom';
import './index.css';
import reportWebVitals from './reportWebVitals';
import HelloErrorBoundaries from "./components/HelloErrorBoundaries";
ReactDOM.createRoot(document.getElementById('root')).render(
    <React.StrictMode>
        <HelloErrorBoundaries/>
    </React.StrictMode>
);
reportWebVitals();
```

9.11.4 运行项目 reactjsbook

使用 npm start 命令运行项目 reactjsbook，在浏览器中输入 localhost:3000，部分效果如图 9-11 所示。

图 9-11 运行项目后在浏览器中输入 localhost:3000 的部分效果

9.12 测试

9.12.1 说明

对于每个测试，通常希望将 React 树渲染给 DOM 元素，以便接收 DOM 事件。当测试结束时，需要清理并卸载树。即使测试失败也需要执行清理；否则，测试可能会导致内存泄漏，并且一个测试可能会影响另一个测试的行为。

可以使用不同的测试模式。常见的测试方法是使用一对 beforeEach 和 afterEach 块，以便组件一直运行，并隔离测试本身造成的影响。在编写 UI 界面测试代码时，可以将渲染、用户事件或数据获取等任务视为与 UI 界面交互的单元。React 提供了 act()方法，它确保在进行任何断言之前，与这些单元相关的所有更新都已处理并应用于 DOM。这有助于使测试运行更接近真实用户在使用应用程序时的体验。要想测试组件对于给定的 props 渲染是否正确，应考虑实现基于 props 渲染消息的简单组件。

有些模块可能在测试环境中不能很好地工作，或者对测试本身不是很重要。使用虚拟数据来 mock（模拟）这些模块，可以使为代码编写测试变得更容易。使用假的 mock 数据进行测试可以防止由于后端不可用而导致的测试不稳定，并使它们运行得更快。而使用一个端到端的框架来运行测试子集，可以显示整个应用程序能否一起工作。

9.12.2 测试简单示例

在 reactjsbook\src\components 目录下创建文件 HelloTest.js，代码如例 9-33 所示。

【例 9-33】 文件 HelloTest.js 的代码。

```
import React from "react";
```

```js
export default function HelloTest(props) {
    if (props.name) {
        return <h1>你好，{props.name}! </h1>;
    } else {
        return <span>嘿，陌生人</span>;
    }
}
```

在 reactjsbook\src\components 目录下创建文件 HelloTest.test.js，代码如例 9-34 所示。

【例 9-34】 文件 HelloTest.test.js 的代码。

```js
import React from "react";
import { render, unmountComponentAtNode } from "react-dom";
import { act } from "react-dom/test-utils";
import HelloTest from "./HelloTest";
let container = null;
beforeEach(() => {
    // 创建一个 DOM 元素作为渲染目标
    container = document.createElement("div");
    document.body.appendChild(container);
});
afterEach(() => {
    // 退出时进行清理
    unmountComponentAtNode(container);
    container.remove();
    container = null;
});
it("渲染有或无名称", () => {
    act(() => {
        render(<HelloTest />, container);
    });
    expect(container.textContent).toBe("嘿，陌生人");
    act(() => {
        render(<HelloTest name="Jenny" />, container);
    });
    expect(container.textContent).toBe("你好，Jenny! ");
    act(() => {
        render(<HelloTest name="Margaret" />, container);
    });
    expect(container.textContent).toBe("你好，Margaret! ");
});
```

右击文件名 HelloTest.test.js，在弹出的快捷菜单中选择 Run 'HelloTest.test.js'菜单项（这些操作统称为测试运行文件 HelloTest.test.js），操作界面如图 9-12 所示。测试运行完之后，效果如图 9-13 所示，显示测试通过。

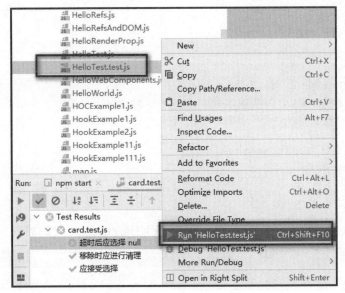

图 9-12　测试运行文件 HelloTest.test.js 的界面

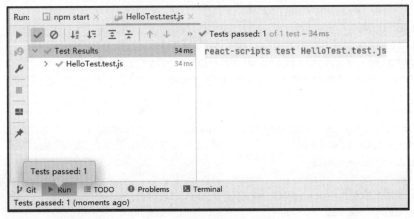

图 9-13　测试通过的效果

9.12.3　异步测试示例

在 reactjsbook\src\components 目录下创建文件 User.js，代码如例 9-35 所示。

【例 9-35】　文件 User.js 的代码。

```
import React, { useState, useEffect } from "react";
export default function User(props) {
   const [user, setUser] = useState(null);
   async function fetchUserData(id) {
      const response = await fetch("/" + id);
      setUser(await response.json());
   }
   useEffect(() => {
      fetchUserData(props.id);
   }, [props.id]);
```

```
    if (!user) {
        return "加载中...";
    }
    return (
        <details>
            <summary>{user.name}</summary>
            <strong>{user.age}</strong> 岁
            <br />
            住在 {user.address}
        </details>
    );
}
```

在 reactjsbook\src\components 目录下创建文件 User.test.js，代码如例 9-36 所示。

【例 9-36】 文件 User.test.js 的代码。

```
import React from "react";
import { render, unmountComponentAtNode } from "react-dom";
import { act } from "react-dom/test-utils";
import User from "./User";
let container = null;
beforeEach(() => {
    container = document.createElement("div");
    document.body.appendChild(container);
});
afterEach(() => {
    unmountComponentAtNode(container);
    container.remove();
    container = null;
});
it("渲染用户数据", async () => {
    const fakeUser = {
        name: "Joni Baez",
        age: "32",
        address: "123, Charming Avenue"
    };
    jest.spyOn(global, "fetch").mockImplementation(() =>
        Promise.resolve({
            json: () => Promise.resolve(fakeUser)
        })
    );
    // 使用异步的 act()方法应用执行成功的 promise
    await act(async () => {
        render(<User id="123" />, container);
    });
    expect(container.querySelector("summary").textContent).toBe(fakeUser.name);
    expect(container.querySelector("strong").textContent).toBe(fakeUser.age);
```

```
        expect(container.textContent).toContain(fakeUser.address);
        // 清理 mock，以确保测试完全隔离
        global.fetch.mockRestore();
});
```

测试运行文件 User.test.js，结果显示测试通过。

9.12.4 mock 测试示例

在 reactjsbook\src\components 目录下创建文件 map.js，代码如例 9-37 所示。

【例 9-37】 文件 map.js 的代码。

```
import React from "react";
import { LoadScript, GoogleMap } from "react-google-maps";
                                            //安装 react-google-maps
export default function Map(props) {
  return (
    <LoadScript id="script-loader" googleMapsApiKey="YOUR_API_KEY">
      <GoogleMap id="example-map" center={props.center} />
    </LoadScript>
  );
}
```

在 reactjsbook\src\components 目录下创建文件 contact.js，代码如例 9-38 所示。

【例 9-38】 文件 contact.js 的代码。

```
import React from "react";
import Map from "./map";
export default function Contact(props) {
  return (
    <div>
      <address>
        联系 {props.name}，通过{" "}
        <a data-testid="email" href={"mailto:" + props.email}>
          email
        </a>
        或者他们的 <a data-testid="site" href={props.site}>
        网站
        </a>。
      </address>
      <Map center={props.center} />
    </div>
  );
}
```

在 reactjsbook\src\components 目录下创建文件 contact.test.js，代码如例 9-39 所示。

【例 9-39】 文件 contact.test.js 的代码。

```
import React from "react";
```

```js
import { render, unmountComponentAtNode } from "react-dom";
import { act } from "react-dom/test-utils";
import Contact from "./contact";
jest.mock("./map", () => {
    return function DummyMap(props) {
        return (
            <div data-testid="map">
                {props.center.lat}:{props.center.long}
            </div>
        );
    };
});
let container = null;
beforeEach(() => {
    container = document.createElement("div");
    document.body.appendChild(container);
});
afterEach(() => {
    unmountComponentAtNode(container);
    container.remove();
    container = null;
});
it("应渲染联系信息", () => {
    const center = { lat: 0, long: 0 };
    act(() => {
        render(
            <Contact
                name="Joni Baez"
                email="test@example.com"
                site="http://test.com"
                center={center}
            />,
            container
        );
    });
    expect(
        container.querySelector("[data-testid='email']").getAttribute("href")
    ).toEqual("mailto:test@example.com");
    expect(
        container.querySelector('[data-testid="site"]').getAttribute("href")
    ).toEqual("http://test.com");
    expect(container.querySelector('[data-testid="map"]').textContent).toEqual(
        "0:0"
    );
});
```

测试运行文件 contact.test.js，结果显示测试通过。

9.12.5 事件测试示例

在 reactjsbook\src\components 目录下创建文件 toggle.js，代码如例 9-40 所示。

【例 9-40】 文件 toggle.js 的代码。

```
import React, { useState } from "react";
export default function Toggle(props) {
    const [state, setState] = useState(false);
    return (
      <button
          onClick={() => {
              setState(previousState => !previousState);
              props.onChange(!state);
          }}
          data-testid="toggle"
      >
          {state === true ? "Turn off" : "Turn on"}
      </button>
    );
}
```

在 reactjsbook\src\components 目录下创建文件 toggle.test.js，代码如例 9-41 所示。

【例 9-41】 文件 toggle.test.js 的代码。

```
import React from "react";
import { render, unmountComponentAtNode } from "react-dom";
import { act } from "react-dom/test-utils";
import Toggle from "./toggle";
let container = null;
beforeEach(() => {
    container = document.createElement("div");
    // container 必须附加到 document，事件才能正常工作
    document.body.appendChild(container);
});
afterEach(() => {
    unmountComponentAtNode(container);
    container.remove();
    container = null;
});
it("点击时更新值", () => {
    const onChange = jest.fn();
    act(() => {
        render(<Toggle onChange={onChange} />, container);
    });
    // 获取按钮元素，并触发点击事件
    const button = document.querySelector("[data-testid=toggle]");
    expect(button.innerHTML).toBe("Turn on");
    act(() => {
```

```
            button.dispatchEvent(new MouseEvent("click", { bubbles: true }));
        });
        expect(onChange).toHaveBeenCalledTimes(1);
        expect(button.innerHTML).toBe("Turn off");
        act(() => {
            for (let i = 0; i < 5; i++) {
                button.dispatchEvent(new MouseEvent("click", { bubbles: true }));
            }
        });
        expect(onChange).toHaveBeenCalledTimes(6);
        expect(button.innerHTML).toBe("Turn on");
    });
```

测试运行文件 toggle.test.js,结果显示测试通过。

习题 9

一、简答题

1. 简述对选择 React 的理解。
2. 简述对 JSX 表示对象的理解。
3. 简述对类组件的执行顺序的理解。
4. 简述对异步编程的理解。
5. 简述对 Fiber 的理解。
6. 简述对模块的理解。
7. 简述对片段的理解。
8. 简述对 context 的理解。
9. 简述对高阶组件的理解。
10. 简述对 ref 转发的理解。
11. 简述对 portal 的理解。
12. 简述对 ref 和 DOM 的理解。
13. 简述对 Web Component 的理解。
14. 简述对 render props 的理解。
15. 简述对错误边界的理解。
16. 简述对测试的理解。

二、实验题

1. 完成片段的应用开发。
2. 完成 context 的应用开发。
3. 完成高阶组件的应用开发。
4. 完成 ref 转发的应用开发。
5. 完成 portal 的应用开发。

6. 完成 ref 和 DOM 的应用开发。
7. 完成 Web Component 的应用开发。
8. 完成 render props 的应用开发。
9. 完成错误边界的应用开发。
10. 完成测试的应用开发。

第 10 章

React应用开发的工具

本章介绍 React 应用开发时用到的包管理器、安装 React、React 开发时用到的编译器和编辑器、构建工具、服务器端渲染工具和 React Router 等内容。深入使用这些工具会涉及较多内容，可以参考这些工具的官方文档。

10.1 包管理器

10.1.1 NPM

NPM（Node.js Package Manager，Node.js 包管理器）是随同 Node.js 一起安装的包管理工具。通过它可以从 NPM 服务器下载第三方包（或下载并安装命令行程序）到本地使用，也支持开发人员将自己编写的包或命令行程序上传到 NPM 服务器供别人使用。可以用命令 npm -v 来测试 NPM 是否安装成功，出现 NPM 版本信息表示安装成功。

NPM 安装包（或命令行程序）是从国外服务器上传、下载的，为了提高上传、下载的速度，可以安装 CNPM 代替 NPM 来上传、下载包（或命令行程序）。

使用 NPM 安装包 packagename 的命令如例 10-1 所示，命令中的 packagename 代表的是要安装的包。

【例 10-1】 使用 NPM 安装包 packagename 的命令。

```
npm install packagename
```

使用 NPM 删除包 packagename 的命令如例 10-2 所示，命令中的 packagename 代表的是要删除的包。

【例 10-2】 使用 NPM 删除包 packagename 的命令。

```
npm remove packagename
```

10.1.2 Yarn

Yarn 是 Facebook、Google 等公司开发的用于替换 NPM 的包管理器。Facebook 已经在生产环境中使用 Yarn，而且 React、React Native、Create React App 等项目都集成了 Yarn。使用 Yarn 时要包含所有共享的代码以及一个描述包的 package.json 文件（称为清单）。

Yarn 允许开发人员共享代码。Yarn 可以快速、安全、可靠地被执行。Yarn 允许开发人员使用其他开发人员的包来解决不同的问题，更轻松地开发软件。Yarn 缓存了曾经下载过的包，所以速度快。在执行代码之前，Yarn 会通过算法校验每个安装包的完整性。

使用 Yarn 安装包 packagename 的命令如例 10-3 所示，命令中的 packagename 代表的是要安装的包。

【例 10-3】 使用 Yarn 安装包 packagename 的命令。

```
yarn add packagename
```

使用 Yarn 删除包 packagename 的命令如例 10-4 所示，命令中的 packagename 代表的是要删除的包。

【例 10-4】 使用 Yarn 删除包 packagename 的命令。

```
yarn remove packagename
```

10.2 安装 React

10.2.1 CDN 链接

可以通过 CDN（内容分发网络）获得 React 和 ReactDOM 的 UMD 版本，如例 10-5 所示。请对照例 1-1 和例 10-5 加深对 CDN 应用的理解。

【例 10-5】 通过 CDN 获得开发环境 React 和 ReactDOM 的代码。

```
<script crossorigin src="https://unpkg.com/react@18.2.0/umd/react.development.js"></script>
<script crossorigin src="https://unpkg.com/react-dom@18.2.0/umd/react-dom.development.js"></script>
```

例 10-5 所示中的 React 和 ReactDOM 版本仅用于开发环境，不适合用于生产环境。应用程序开发完成并经过压缩优化后可通过如例 10-6 所示的方式引用适合于生产的 React 版本。

【例 10-6】 通过 CDN 获得生产环境 React 和 ReactDOM 的代码。

```
<script crossorigin src="https://unpkg.com/react@18.2.0/umd/react.production.min.js"></script>
<script crossorigin src="https://unpkg.com/react-dom@18.2.0/umd/react-dom.production.min.js"></script>
```

通过 CDN 的方式引入 React，可以设置 crossorigin 属性，如例 10-5 所示。这样能在

React 16.0 及以上的版本中有更好的错误处理体验。

10.2.2　Create React App

Create React App 是 React 官方推出的一个开发 React 应用程序的脚手架（命令行工具），其是用 React 创建应用程序的较好方法。

使用 Create React App 可以配置开发环境，使用最新 JavaScript 特性，提供良好的开发体验，为生产环境优化应用程序。这需要安装 14.0 以上版本的 Node.js 和 5.6 以上版本的 NPM。可以使用如例 10-7 所示的命令创建项目 myapp。

【例 10-7】　创建项目 myapp 的命令。

```
npx create-react-app myapp
cd myapp
npm start
```

例 10-7 所示中的 npx 是 5.2 以上版本 NPM 自带的包执行器。npx 要解决的主要问题是调用项目内部安装的模块，它可以让 NPM 包中的命令行工具和其他可执行文件的使用变得更加简单。

Create React App 不会处理后端逻辑或操纵数据库，但是可以使用它来配合任何技术栈使用的后端。它通过 react-scripts 在内部集成了 Babel、webpack、ESLint，并配置了一系列内置的 loader 和默认的 NPM 的脚本，可以很轻松地实现零配置（即开箱即用）就可以快速开发 React 的应用程序。使用 WebStorm 等集成开发环境创建 React 项目时也使用默认的脚手架 Create React App 来创建项目，如图 8-1 所示。

10.3　编译器和编辑器

10.3.1　Babel

Babel 是一个编译器，主要用于将采用 ES 6 以上版本语法编写的代码转换为向后兼容的 JavaScript 语法，以便能够运行在当前和旧版本的浏览器或其他环境中。Babel 可以进行语法转换、源码转换，可以通过 polyfill 方式在目标环境中添加缺失的特性。Babel 通过语法转换器来支持新版本的 JavaScript 语法。Babel 能够转换 JSX 语法。Bebel 通过预设（preset）设置执行什么转换。

10.3.2　ESLint

Lint 是代码格式检查工具的一个统称，包括 JSLint、ESLint 等工具。ESLint 是一个使用 Node.js 编写的开源 JavaScript 代码检查工具。代码检查是一种静态的分析，常用于寻找有问题的模式或者代码，并且不依赖于具体的编码风格。对于大多数编程语言来说都会有代码检查工具，一般的编译程序会内置检查工具。像 ESLint 这样的工具可以帮助开发人员在编码的过程中发现问题，而不是等到应用程序执行时（运行时）发现问题。

为了便于人们使用，ESLint 内置了一些代码检查规则，可以在使用过程中自定义代码检查规则。ESLint 的所有代码检查规则都被设计成可插入的。ESLint 的默认规则与其他的插件并没有什么区别，规则本身和测试可以依赖于同样的模式。

10.3.3　Prettier

Prettier 是一个代码格式化工具，能够解析代码，可以使用自己设定的规则来重新打印出格式规范的代码。它具有可配置、支持多种语言、集成多数编辑器、配置简洁等优点。在代码评审时使用 Prettier 不需要再讨论代码样式，节省了时间与精力。Prettier 可以和 ESLint 集成使用。其中，ESLint 负责保证代码质量，Prettier 负责处理代码格式化。

10.3.4　PropTypes

可以使用 PropTypes 进行类型检查来找出更多的 bug。对于某些应用程序，也可以使用 JavaScript 扩展（如 Flow 或 TypeScript）对整个应用程序进行类型检查。即使不使用这些工具，React 也有一些内置的类型检查能力。

10.4　构建工具

10.4.1　webpack

webpack 是一个用于 JavaScript 应用程序的静态模块打包器（bundler）。webpack 处理应用程序时，会将项目中所需的每一个模块组合成一个或多个 bundle，它们均为静态资源，用于展示内容。从 4.0 版开始，webpack 可以不用再引入一个配置文件（而使用默认设置）来打包项目。

每当一个文件依赖（即用到）另一个文件时，webpack 都会将两者之间视为存在依赖关系。当 webpack 处理应用程序时，它会根据命令行参数中或配置文件中定义的模块列表开始处理。从入口资源（文件）开始，webpack 将沿着导入树，将用到的所有模块都放入构建包中。遍历这些文件可以得到一个依赖图，这个依赖图包含着应用程序中所需的每个模块，然后将所有模块打包为少量的 bundle，通常只有一个 bundle 可由浏览器加载。

webpack 作为模块打包工具，它的优势体现在模块化、组合 React 组件构成应用程序和提升网络性能。除了打包之外，webpack 还可以进行代码分拆、代码简化、特性标签、模块热替换（更新有变化的模块）等功能。

10.4.2　Parcel

Parcel 是应用程序打包工具。它利用多核处理提供了极快的速度，并且不需要任何配置。Parcel 可以使用任何类型的文件作为入口，但是最好还是使用 HTML 或 JavaScript 文件。如果在 HTML 中使用相对路径引入主要的 JavaScript 文件，Parcel 也将会对它进行处理，将其替换为相对于输出文件的 URL 地址。

Parcel 内置了一个改变文件时能够自动重新构建应用程序的开发服务器，而且为了实现快速开发，该开发服务器支持模块热替换。

10.5 服务器端渲染工具

10.5.1 Next.js

Next.js 是一个流行的轻量级框架,用于配合 React 打造静态化和服务端渲染应用程序。它包括开箱即用的样式和路由方案,并且假定使用 Node.js 作为服务器环境。

在 Next.js 中,一个页面就是一个从 .js、.jsx、.ts 或 .tsx 文件导出的 React 组件,这些文件存放在 pages 目录下。每个页面都使用其文件名作为路由。如果在 pages 目录下创建了一个命名为 about.js 的文件并导出一个如例 10-8 所示的 React 组件 About,就可以通过 /about 路径进行访问。

【例 10-8】 组件 About 的代码。

```
function About() {
  return <div>About</div>
}
export default About
```

Next.js 支持具有动态路由的页面。例如,如果在 pages\posts 目录下创建了 1.js 和 2.js 两个文件,就可以通过如 posts/[id](代表 posts/1、posts/2 等)所示的路径进行访问。

通常,Next.js 会预渲染每个页面。这意味着 Next.js 会预先为每个页面生成 HTML 文件,而不是由客户端 JavaScript 来完成。预渲染可以带来更好的性能和 SEO 效果。Next.js 具有静态生成(推荐使用,性能更好)、服务器端预渲染两种形式的预渲染。这两种方式的不同之处在于生成 HTML 页面的时间不同。静态生成预渲染的 HTML 在构建时生成,并在每次页面请求时重用。服务器端预渲染是在每次页面请求时重新生成 HTML。Next.js 允许为每个页面选择预渲染方式,即可以创建一个混合预渲染的 Next.js 应用程序(如对大多数页面使用静态生成预渲染的方式,对其他页面使用服务器端预渲染方式)。

如果一个页面使用了静态生成预渲染的方式,在构建时将生成页面对应的 HTML 文件。这意味着,在生产环境中,运行 next build 时将生成该页面对应的 HTML 文件。然后,此 HTML 文件将在每个页面请求时被重用,还可以被 CDN 缓存。

在某些情况下,服务器端预渲染可能是唯一的选择。使用服务器端渲染(SSR,也称为动态渲染)时,Next.js 针对每个页面的请求进行预渲染,可以保证页面始终是最新的。由于 CDN 无法缓存该页面,因此速度会较慢。于是,可以将静态生成预渲染与客户端渲染一起使用,可以跳过页面某些部分的预渲染,然后使用客户端 JavaScript 来填充它们。

10.5.2 Razzle

Razzle 是一个无须配置的服务端渲染框架,用于在服务端渲染 React 应用程序,但它比 Next.js 更灵活。可以通过它将一般的 JavaScript 应用程序抽象成单个的依赖,然后将框架、路由和数据提取出来。同时,Razzle 支持可插拔渲染。它具有通用的热更新模块,因此当修改时,客户端和服务器都会进行更新,不需要进行重启。它支持渲染 React、Angular

和 Vue 等框架，与 Create React App 有着相同的 CSS 设置，提供了 ES 6 语法糖，支持使用 Jest 进行测试。

10.5.3 Gatsby

Gatsby 是一个基于 React 的免费开源框架，用于帮助开发人员实现运行速度极快的应用程序。Gatsby 是用 React 创建静态网站的较好方式之一。Gatsby 可以创建的项目类型很多，通常用于构建内容驱动型网站。

10.6 React Router

视频讲解

10.6.1 说明

React Router 是一个基于 React 之上的强大路由库，利用它可以向应用程序中快速地添加视图和数据流，同时保持页面与 URL 间的同步。有关 React Router 的更多内容请参考 React Router 官方文档。

10.6.2 创建组件

在项目 reactjsbook 的 src 目录下创建 routerdemo 子目录，在 reactjsbook\src\ routerdemo 目录下创建文件 App.js，代码如例 10-9 所示。

【例 10-9】文件 App.js 的代码。

```
import React from "react";
import {BrowserRouter as Router, Route, Routes} from "react-router-dom";
import {
    Home,
    ChildrenEx,
    Services,
    History,
    LinkEx,
    AHrefEx
} from "./page";
function App() {
    return(
        <Router>
            <Routes>
                <Route exact path="/" element={<Home/>} />
                <Route exact path="/childrenEx" element={<ChildrenEx/>} />
                <Route exact path="/childrenEx/services" element=
{<Services/>} />
                <Route exact path="/childrenEx/history" element={<History/>}/>
                <Route exact path="/linkEx" element={<LinkEx/>} />
                <Route exact path="/aHrefEx" element={<AHrefEx/>} />
            </Routes>
```

```
        </Router>
    )
}
export default App;
```

在 reactjsbook\src\routerdemo 目录下创建文件 page.js,代码如例 10-10 所示。

【例 10-10】 文件 page.js 的代码。

```
import React from "react";
import {Link} from "react-router-dom";
export function Home() {
    return (
        <div>
            <h1>React Router 示例</h1>
            <a href="/childrenEx">有下级(二级)路由的示例</a><br/>
            <Link to="linkEx">没有下级路由,使用 Link 的示例</Link><br/>
            <a href="/aHrefEx">没有下级路由,使用超链接的示例</a><br/>
        </div>
    );
}
export function ChildrenEx() {
    return (
        <div>
            <h1>有下级(二级)路由的示例</h1>
            <Link to="history">路由</Link><br/>
            <Link to="services">组件</Link>
        </div>
    );
}
export function LinkEx() {
    return (
        <div>
            <h1>没有下级路由,使用 Link 的示例</h1>
        </div>
    );
}
export function AHrefEx() {
    return (
        <div>
            <h1>没有下级路由,使用超链接的示例</h1>
        </div>
    );
}
export function Services() {
    return (
        <section>
            <h2>组件</h2>
            <p>
                多个组件保存在一起,在文件 page.js 中。
```

```
            </p>
        </section>
    );
}
export function History() {
    return (
        <section>
            <h2>路由</h2>
            <p>
                路由信息保存在一起，在文件 App.js 中。
            </p>
        </section>
    );
}
```

10.6.3　修改文件 index.js

修改 reactjsbook\src 目录下文件 index.js 的代码，如例 10-11 所示。

【例 10-11】　修改后的文件 index.js 的代码。

```
import React from 'react';
import ReactDOM from 'react-dom';
import './index.css';
import reportWebVitals from './reportWebVitals';
import App from "./routerdemo/App";
ReactDOM.createRoot(document.getElementById('root')).render(
  <React.StrictMode>
    <App/>
  </React.StrictMode>
);
reportWebVitals();
```

10.6.4　运行项目 reactjsbook

使用 npm start 命令运行项目 reactjsbook，在浏览器中输入 localhost:3000，效果如图 10-1 所示。单击图 10-1 所示中的超链接"有下级（二级）路由的示例"的效果如图 10-2 所示。

图 10-1　运行项目后在浏览器中输入 localhost:3000 的效果

图 10-2　单击图 10-1 所示中的超链接"有下级（二级）路由的示例"的效果

单击图10-2所示中的超链接"路由"的效果如图10-3所示。单击图10-2所示中的超链接"组件"的效果如图10-4所示。

图10-3　单击图10-2所示中的超链接"路由"的效果

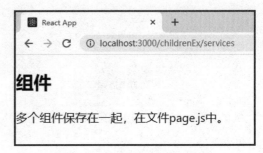

图10-4　单击图10-2所示中的超链接"组件"的效果

习题 10

一、简答题

1. 简述对 webpack 的理解。
2. 简述对 Next.js 的理解。

二、实验题

1. 用 NPM 安装包。
2. 用 Yarn 安装包。
3. 完成 CDN 链接的 React 应用开发。
4. 用 Create React App 创建 React 项目。
5. 完成 Babel 的应用开发。
6. 完成 React Router 的应用开发。

第三部分 实战篇

第 11 章　React 与 Redux 的整合开发
第 12 章　React 与 Spring Boot 的整合开发
第 13 章　React 与 Python 框架的整合开发
第 14 章　React 与 Go 的整合开发
第 15 章　案例——实现一个简易的员工信息管理系统

第 11 章

React与Redux的整合开发

本章先简要介绍 Redux 的动机、核心内容、React 与 Redux 对比，再介绍用 React 和 Redux 整合开发计数器、待办事项管理小工具等内容。

11.1 React 与 Redux 概述

11.1.1 Redux 动机

随着应用程序开发日趋复杂，JavaScript 需要管理很多的 state（状态或称为状态取值）。这些 state 可能包括服务器响应、缓存数据、本地生成尚未持久化到服务器的数据、激活的路由、被选中的标签、分页器等。由于可能无法较好地跟踪和管理 state 在什么时候、由于什么原因发生变化等内容（即 state 可能不受控制），管理不断增多和变化的 state 将变得非常困难。当应用程序代码比较复杂时，想重现问题或者添加新功能就会变得十分困难。

新需求（如更新调优、渲染服务端、路由跳转前请求数据等）使得前端开发正在经受前所未有的复杂性挑战。一些库（如 React）试图通过在视图层禁止异步编程和禁止直接操作 DOM 等手段，解决这个问题。但是，在进行 React 应用开发时，开发人员依然需要处理 state 中数据。这时，可以借助 Redux 来更好地完成这项工作。

11.1.2 Redux 核心内容

Redux 的核心内容包括 store（存储）、action、reducer。store 是用来存储应用程序 state 树的一个 object 对象，state 树包含了应用程序所有的 state。改变 store 内 state 的唯一途径是对它派发（dispatch）一个 action。

当使用普通对象来描述应用程序的 state 时，state 代码如例 11-1 所示。

【例 11-1】state 的代码。

```
{
  todos: [{
    text: 'Eat food',
    completed: true
  }, {
    text: 'Exercise',
    completed: false
  }],
  visibilityFilter: 'SHOW_COMPLETED'
}
```

例 11-1 所示中的这个对象就像模型（model），区别是它并没有 setter()方法。因此其他的代码不能随意修改它，以避免造成难以复现的 bug。

要想更新 state 中的数据，需要发起一个 action。action 表示应用程序中的各类动作或操作，不同的操作会改变应用相应的 state，action 是一个带 type 属性的对象。

action 就是一个普通 JavaScript 对象用来描述发生了什么。action 的代码如例 11-2 所示。

【例 11-2】 action 的代码。

```
{ type: 'ADD_TODO', text: 'Go to swimming pool' }
{ type: 'TOGGLE_TODO', index: 1 }
{ type: 'SET_VISIBILITY_FILTER', filter: 'SHOW_ALL' }
```

强制使用 action 来描述所有变化带来的好处是可以清晰地知道应用程序中到底发生了什么。如果一些东西改变了，还可以知道为什么变。

为了把 action 和 state 串起来，需要开发一些函数，这就是 reducer。reducer 是一个包含 state 和 action 两个参数并返回新 state 的函数。它可以被用来初始化、计算并返回新的 state。reducer 不允许有副作用，不能操作 DOM，不能发 Ajax 请求，更不能直接修改 state。对于大的应用程序来说，不大可能仅仅只编写一个这样的函数，所以可能需要编写很多小函数来分别管理 state 的一部分数据。reducer 的代码如例 11-3 所示。

【例 11-3】 reducer 的代码。

```
function visibilityFilter(state = 'SHOW_ALL', action) {
  if (action.type === 'SET_VISIBILITY_FILTER') {
    return action.filter;
  } else {
    return state;
  }
}
function todos(state = [], action) {
  switch (action.type) {
  case 'ADD_TODO':
    return state.concat([{ text: action.text, completed: false }]);
  case 'TOGGLE_TODO':
    return state.map((todo, index) =>
      action.index === index ?
```

```
            { text: todo.text, completed: !todo.completed } :
            todo
      )
    default:
      return state;
  }
}
```

可以再开发一个 reducer 调用例 11-3 所示代码中的两个 reducer，进而来管理（整合）整个应用程序的 state。整合 reducer 代码如例 11-4 所示。

【例 11-4】 整合 reducer 的代码。

```
function todoApp(state = {}, action) {
  return {
    todos: todos(state.todos, action),
    visibilityFilter: visibilityFilter(state.visibilityFilter, action)
  };
}
```

11.1.3 React 与 Redux 对比

Redux 支持 React、Angular、Ember、jQuery 甚至纯 JavaScript。尽管如此，Redux 还是和 React、Deku 这些库搭配起来用最好，因为这些库允许以 state()函数的形式来描述 UI 界面，Redux 通过 action 的形式来发起 state 变化。

Redux 应用开发时往往需要引入 redux 和 react-redux 两个库。Redux 是用来管理数据的架构模式，它不关心用的是什么框架（库），如 React、Vue。react-redux 是把 Redux 和 React 结合起来的库。

无 state 组件负责渲染以及用户交互，它只有 props。store 的 state 通过容器组件注入无 state 组件，引起重新渲染。容器组件可以派发 action，引起 store 的 state 改变。

11.2 计数器的开发

11.2.1 创建 action

在项目 reactjsbook\src 目录下创建 counter 子目录，在 reactjsbook\src\counter 目录下创建 actions 子目录，再在 reactjsbook\src\counter\actions 目录下创建文件 Action.js，代码如例 11-5 所示。

【例 11-5】 文件 Action.js 的代码。

```
export const INCREASE='INCREASE';
export const DECREASE='DECREASE';
export function incActionGenerator(){
  return {
    type:INCREASE,
```

```
            preload:'increase the current value'
        }
    }
    export function decActionGenerator(){
        return {
            type:DECREASE,
            preload:'decrease the current value'
        }
    }
```

11.2.2 创建 reducer

在项目 reactjsbook\src\counter 目录下创建 reducers 子目录，再在 reactjsbook\src\counter\reducers 目录下创建文件 Reducer.js，代码如例 11-6 所示。

【例 11-6】 文件 Reducer.js 的代码。

```
import {
    INCREASE, DECREASE
} from "../actions/Action";
const changeValue=(state={title:'React+Redux 实现简易计数器',info:'数值为：',value:0},action)=>{
    switch(action.type){
        case INCREASE:
            return {...state,value:state.value+1};
        //或 return {value:state.value+1};
        case DECREASE:
            return {...state,value:state.value-1};
        default:
            return state;
    }
}
export default changeValue;
```

11.2.3 创建组件

在项目 reactjsbook\src\counter 目录下创建 components 子目录，再在 reactjsbook\src\counter\components 目录下创建文件 Counter.js，代码如例 11-7 所示。

【例 11-7】 文件 Counter.js 的代码。

```
import React from "react";
export default class Counter extends React.Component {
    render() {
        const {title,info,value,onIncClick,onDecClick}=this.props;
        return (
            <div>
                <h1>{title}</h1>
                <span>{info}{value}</span>
                <br />
```

```
                <button type="button" onClick={onIncClick}>加一</button>
                <button type="button" onClick={onDecClick}>减一</button>
            </div>
        )
    }
}
```

在项目 reactjsbook\src\counter 目录下创建 containers 子目录，再在 reactjsbook\src\counter\containers 目录下创建文件 CounterContainer.js，代码如例 11-8 所示。

【例 11-8】 文件 CounterContainer.js 的代码。

```
import {connect} from 'react-redux'    //安装包 react-redux
import Counter from '../components/Counter'
import {incActionGenerator,decActionGenerator} from "../actions/Action";
const mapStateToProps=(state)=>({
    title:state.title,
    info:state.info,
    value:state.value
})
const mapDispatchToProps=(dispatch)=>({
    onIncClick:()=>dispatch(incActionGenerator()),
    onDecClick:()=>dispatch(decActionGenerator())
})
const MyCounterApp=connect(
    mapStateToProps,
    mapDispatchToProps
)(Counter);
export default MyCounterApp;
```

11.2.4　修改文件 index.js

修改 reactjsbook\src 目录下文件 index.js 的代码，如例 11-9 所示。

【例 11-9】 修改后的文件 index.js 的代码。

```
import React from 'react';
import ReactDOM from 'react-dom';
import { createStore } from 'redux';   //安装包 redux
import './index.css';
import reportWebVitals from './reportWebVitals';
import {Provider} from 'react-redux';
import MyCounterApp from "./counter/containers/CounterContainer";
import changeValue from './counter/reducers/Reducer'
const store = createStore(changeValue);//用 Reducer 来创建 store
ReactDOM.createRoot(document.getElementById('root')).render(
    <React.StrictMode>
        <Provider store={store}>
            <MyCounterApp/>
        </Provider>
```

```
        </React.StrictMode>
);
reportWebVitals();
```

11.2.5 运行项目 reactjsbook

使用 npm start 命令运行项目 reactjsbook，在浏览器中输入 localhost:3000，效果如图 11-1 所示。单击图 11-1 所示中的"加一"按钮的效果如图 11-2 所示。单击图 11-2 所示中的"减一"按钮的效果如图 11-1 所示。

图 11-1　运行项目后在浏览器中输入 localhost:3000 的效果

图 11-2　单击图 11-1 所示中的"加一"按钮的效果

11.3 待办事项管理小工具的开发

11.3.1 创建 action

在项目 reactjsbook\src 目录下创建 todolists 子目录，在 reactjsbook\src\todolists 目录下创建 actions 子目录，再在 reactjsbook\src\todolists\actions 目录下创建文件 Action.js，代码如例 11-10 所示。

【例 11-10】文件 Action.js 的代码。

```
let nextTodoId = 0
export const addTodo = text => {
    return {
        type: 'ADD_TODO',
        id: nextTodoId++,
```

```
        text
    }
}
export const setVisibilityFilter = filter => {
    return {
        type: 'SET_VISIBILITY_FILTER',
        filter
    }
}
export const toggleTodo = id => {
    return {
        type: 'TOGGLE_TODO',
        id
    }
}
```

11.3.2 创建 reducer

在项目 reactjsbook\src\todolists 目录下创建 reducers 子目录，再在 reactjsbook\src\todolists\reducers 目录下创建文件 Reducer.js，代码如例 11-11 所示。

【例 11-11】 文件 Reducer.js 的代码。

```
import { combineReducers } from 'redux'
const todos = (state = [], action) => {
    switch (action.type) {
        case 'ADD_TODO':
            return [
                ...state,
                {
                    id: action.id,
                    text: action.text,
                    completed: false
                }
            ]
        case 'TOGGLE_TODO':
            return state.map(todo =>
                (todo.id === action.id)
                    ? {...todo, completed: !todo.completed}
                    : todo
            )
        default:
            return state
    }
}
const visibilityFilter = (state = 'SHOW_ALL', action) => {
    switch (action.type) {
        case 'SET_VISIBILITY_FILTER':
            return action.filter
```

```
            default:
                return state
        }
    }
    const reducer = combineReducers({
        todos,
        visibilityFilter
    })
    export default reducer
```

11.3.3 创建组件

在项目 reactjsbook\src\todolists 目录下创建 components 子目录，再在 reactjsbook\src\todolists\components 目录下创建组件 Todo、TodoList、Link、Footer、App，对应的文件名依次为 Todo.js、TodoList.js、Link.js、Footer.js、App.js，对应的代码依次如例 11-12～例 11-16 所示。

【例 11-12】 文件 Todo.js 的代码。

```
import React from 'react'
import PropTypes from 'prop-types'   //安装包 prop-types
const Todo = ({ onClick, completed, text }) => (
  <li
    onClick={onClick}
    style={{
      textDecoration: completed ? 'line-through' : 'none'
    }}
  >
    {text}
  </li>
)
Todo.propTypes = {
  onClick: PropTypes.func.isRequired,
  completed: PropTypes.bool.isRequired,
  text: PropTypes.string.isRequired
}
export default Todo
```

【例 11-13】 文件 TodoList.js 的代码。

```
import React from 'react'
import PropTypes from 'prop-types'
import Todo from './Todo'
const TodoList = ({ todos, onTodoClick }) => (
  <ul>
    {todos.map(todo => (
      <Todo key={todo.id} {...todo} onClick={() => onTodoClick(todo.id)} />
    ))}
  </ul>
```

```
)
TodoList.propTypes = {
  todos: PropTypes.arrayOf(
    PropTypes.shape({
      id: PropTypes.number.isRequired,
      completed: PropTypes.bool.isRequired,
      text: PropTypes.string.isRequired
    }).isRequired
  ).isRequired,
  onTodoClick: PropTypes.func.isRequired
}
export default TodoList
```

【例 11-14】 文件 Link.js 的代码。

```
import React from 'react'
import PropTypes from 'prop-types'
const Link = ({ active, children, onClick }) => {
  if (active) {
    return <span>{children}</span>
  }
  return (
    <a
      href=""
      onClick={e => {
        e.preventDefault()
        onClick()
      }}
    >
      {children}
    </a>
  )
}
Link.propTypes = {
  active: PropTypes.bool.isRequired,
  children: PropTypes.node.isRequired,
  onClick: PropTypes.func.isRequired
}
export default Link
```

【例 11-15】 文件 Footer.js 的代码。

```
import React from 'react'
import {FilterLink} from "../containers/TodolistsContainer";
const Footer = () => (
  <p>
    显示待办事项情况：
    {' '}
    <FilterLink filter="SHOW_ALL">
      所有事项
```

```
        </FilterLink>
        {', '}
        <FilterLink filter="SHOW_ACTIVE">
            未办理事项
        </FilterLink>
        {', '}
        <FilterLink filter="SHOW_COMPLETED">
            已办结事项
        </FilterLink>
    </p>
)
export default Footer
```

【例 11-16】 文件 App.js 的代码。

```
import React from 'react'
import Footer from './Footer'
import {AddTodo, VisibleTodoList} from "../containers/TodolistsContainer";
const App = () => (
  <div>
    <AddTodo />
    <VisibleTodoList />
    <Footer />
  </div>
)
export default App
```

在项目 reactjsbook\src\todolists 目录下创建 containers 子目录，再在 reactjsbook\src\todolists\containers 目录下创建文件 TodolistsContainer.js，代码如例 11-17 所示。

【例 11-17】 文件 TodolistsContainer.js 的代码。

```
import { connect } from 'react-redux'
import React from 'react'
import TodoList from "../components/TodoList";
import {addTodo, setVisibilityFilter, toggleTodo} from "../actions/Action";
import Link from "../components/Link";
//FilterLink
const mapStateToProps = (state, ownProps) => {
    return {
        active: ownProps.filter === state.visibilityFilter
    }
}
const mapDispatchToProps = (dispatch, ownProps) => {
    return {
        onClick: () => {
            dispatch(setVisibilityFilter(ownProps.filter))
        }
    }
}
```

```jsx
const FilterLink = connect(
    mapStateToProps,
    mapDispatchToProps
)(Link)
//VisibleTodoList
const getVisibleTodos = (todos, filter) => {
    switch (filter) {
        case 'SHOW_COMPLETED':
            return todos.filter(t => t.completed)
        case 'SHOW_ACTIVE':
            return todos.filter(t => !t.completed)
        case 'SHOW_ALL':
        default:
            return todos
    }
}
const mapStateToProps2 = state => {
    return {
        todos: getVisibleTodos(state.todos, state.visibilityFilter)
    }
}
const mapDispatchToProps2 = dispatch => {
    return {
        onTodoClick: id => {
            dispatch(toggleTodo(id))
        }
    }
}
const VisibleTodoList = connect(
    mapStateToProps2,
    mapDispatchToProps2
)(TodoList)
//AddTodo
let AddTodo = ({ dispatch }) => {
    let input
    return (
        <div>
            <form
                onSubmit={e => {
                    e.preventDefault()
                    if (!input.value.trim()) {
                        return
                    }
                    dispatch(addTodo(input.value))
                    input.value = ''
                }}
            >
                <h1>React+Redux 实现待办事项管理小工具</h1>
                <input
```

```
                    ref={node => {
                        input = node
                    }}
                />
                <button type="submit">
                    增加待办事件
                </button>
            </form>
        </div>
    )
}
AddTodo = connect()(AddTodo)
export {FilterLink,VisibleTodoList,AddTodo}
```

11.3.4　修改文件 index.js

修改 reactjsbook\src 目录下文件 index.js 的代码，如例 11-18 所示。

【例 11-18】　修改后的文件 index.js 的代码。

```
import React from 'react';
import ReactDOM from 'react-dom';
import { createStore } from 'redux';
import './index.css';
import reportWebVitals from './reportWebVitals';
import {Provider} from 'react-redux';
import App from "./todolists/components/App";
import reducer from "./todolists/reducers/Reducer";
const store = createStore(reducer)
ReactDOM.createRoot(document.getElementById('root')).render(
    <React.StrictMode>
        <Provider store={store}>
            <App/>
        </Provider>
    </React.StrictMode>
);
reportWebVitals();
```

11.3.5　运行项目 reactjsbook

使用 npm start 命令运行项目 reactjsbook，在浏览器中输入 localhost:3000，效果如图 11-3 所示。在图 11-3 所示的文本框中输入"学习 React"，如图 11-4 所示。单击图 11-4 所示中的"增加待办事项"按钮，效果如图 11-5 所示。单击图 11-5 所示中的超链接"未办理事项"的效果如图 11-6 所示。单击图 11-6 所示中的文本"学习 React"所在区域后单击超链接"已办结事项"，效果如图 11-7 所示，显示"学习 React"事项已经办结。继续增加待办理事项后显示所有事项的效果如图 11-8 所示。

图 11-3　运行项目后在浏览器中输入 localhost:3000 的效果

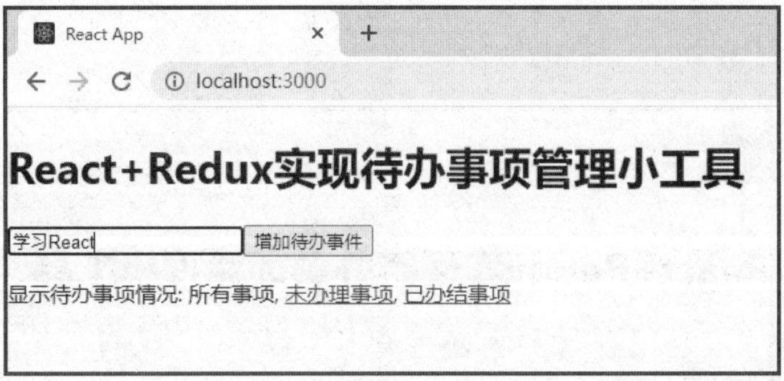

图 11-4　在图 11-3 所示中的文本框中输入"学习 React"的操作界面

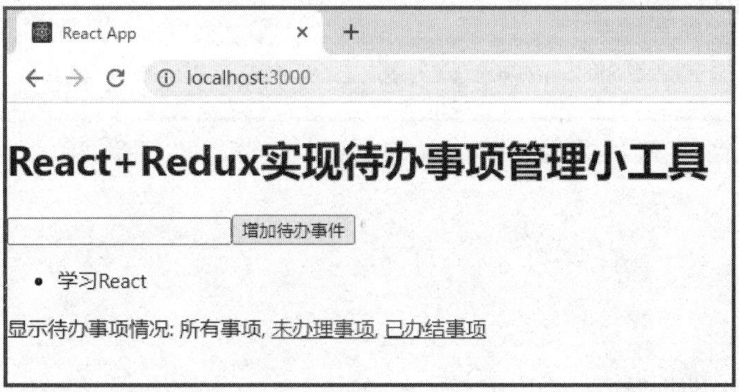

图 11-5　单击图 11-4 所示中的"增加待办事项"按钮的效果

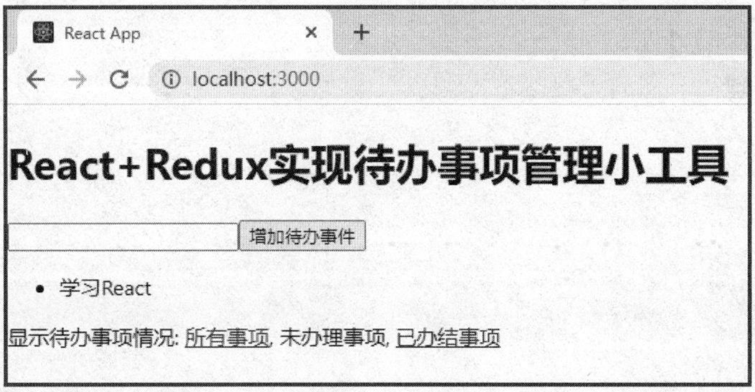

图 11-6　单击图 11-5 所示中的超链接"未办结事项"的效果

图 11-7　单击图 11-6 所示中的文本"学习 React"所在
区域后单击超链接"已办结事项"的效果

图 11-8　继续增加待办理事项后显示所有事项的效果

习题 11

一、简答题

简述对 Redux 的理解。

二、实验题

1. 完成计数器的开发。
2. 完成待办理事项管理小工具的开发。

第12章

React与Spring Boot的整合开发

视频讲解

本章先简要介绍 Spring 的构成和 Spring Boot 的特点，再介绍 React 与 Spring Boot、MySQL 数据库的整合开发示例。此示例演示了 React 用 fetch 方式访问 Spring Boot。

12.1 Spring Boot 简介

12.1.1 Spring 的构成

Spring 提供了非常多的子项目，这些子项目可以帮助开发人员更好地进行相关开发。其中，Spring Framework Core 是 Spring 的核心项目，Spring Framework Core 中包含了一系列 IoC 容器的设计，提供了依赖注入的实现；还集成了面向切面编程（AOP）、MVC、JDBC、事务处理模块的实现。

Spring Boot 提供了快速构建 Spring 应用的方法，达到了"开箱即用"；使用默认的 Java 配置来实现快速开发，并"即时运行"。

Spring Data 提供了对主流关系型数据库提供支持，并提供使用非关系型数据的能力等。

12.1.2 Spring Boot 的特点

Spring Boot 使得开发人员可以很容易地创建基于 Spring 的应用和服务。Spring Boot 的特点包括：

（1）约定大于配置。通过代码结构、注解的约定和命名规范等方式来减少配置，减少冗余代码，使得编码变得更加简单。

（2）内嵌有 Tomcat（或 Netty），打包方式不再强制要求打成 War 包来部署，可以直接采用 Jar 包。

（3）简化 Maven 配置，并推荐使用 Gradle 替代 Maven 进行项目管理。Maven 用于项目的构建，主要可以对依赖包进行管理，Maven 将项目所使用的依赖包信息放到文件 pom.xml 的<dependencies></dependencies>节点之间。

（4）定制"开箱即用"的 Starter，没有代码生成，也无须 XML 配置；还可以修改默认值来满足特定的需求。

（5）提供生产就绪型功能，即提供了一些大型项目中常见的非功能特性。如嵌入式服务器、安全、指标、健康检查、外部配置等内容。

12.2 Spring Boot 作为后端的开发

12.2.1 创建项目 backendforreactjs

参考本书电子版附录 B 或其他资源，用 IDEA 创建项目 backendforreactjs，修改文件 pom.xml，修改后的文件 pom.xml 的代码如例 12-1 所示。

【例 12-1】 修改后的文件 pom.xml 的代码。

```xml
<?xml version="1.0" encoding="UTF-8"?>
<project xmlns="http://maven.apache.org/POM/4.0.0" xmlns:xsi="http://www.w3.org/2001/XMLSchema-instance"
     xsi:schemaLocation="http://maven.apache.org/POM/4.0.0 https://maven.apache.org/xsd/maven-4.0.0.xsd">
    <modelVersion>4.0.0</modelVersion>
    <parent>
        <groupId>org.springframework.boot</groupId>
        <artifactId>spring-boot-starter-parent</artifactId>
        <version>2.5.6</version>
        <relativePath/> <!-- lookup parent from repository -->
    </parent>
    <groupId>com.bookcode</groupId>
    <artifactId>backendforreactjs</artifactId>
    <version>0.0.1-SNAPSHOT</version>
    <name>backendforreactjs</name>
    <description>Demo project for Spring Boot</description>
    <properties>
        <java.version>11</java.version>
    </properties>
    <dependencies>
        <dependency>
            <groupId>org.springframework.boot</groupId>
            <artifactId>spring-boot-starter-data-jpa</artifactId>
        </dependency>
        <dependency>
            <groupId>org.springframework.boot</groupId>
            <artifactId>spring-boot-starter-web</artifactId>
        </dependency>
        <dependency>
```

```xml
            <groupId>mysql</groupId>
            <artifactId>mysql-connector-java</artifactId>
        </dependency>
        <dependency>
            <groupId>org.projectlombok</groupId>
            <artifactId>lombok</artifactId>
            <optional>true</optional>
        </dependency>
        <dependency>
            <groupId>org.springframework.boot</groupId>
            <artifactId>spring-boot-starter-test</artifactId>
            <scope>test</scope>
        </dependency>
    </dependencies>
    <build>
        <plugins>
            <plugin>
                <groupId>org.springframework.boot</groupId>
                <artifactId>spring-boot-maven-plugin</artifactId>
                <configuration>
                    <excludes>
                        <exclude>
                            <groupId>org.projectlombok</groupId>
                            <artifactId>lombok</artifactId>
                        </exclude>
                    </excludes>
                </configuration>
            </plugin>
        </plugins>
    </build>
</project>
```

12.2.2 创建类和接口

创建类 User，代码如例 12-2 所示。

【例 12-2】 类 User 的代码。

```java
package com.bookcode.backendforreactjs;
import javax.persistence.*;
@Entity
@Table(name="user")
public class User {
    @Id
    @GeneratedValue(strategy = GenerationType.IDENTITY)
    private Integer uid;
    private String name;
    private String age;
    public User( ) {
```

```java
        }
        public Integer getUid() {
            return uid;
        }
        public void setUid(Integer uid) {
            this.uid = uid;
        }
        public String getName() {
            return name;
        }
        public void setName(String name) {
            this.name = name;
        }
        public String getAge() {
            return age;
        }
        public void setAge(String age) {
            this.age = age;
        }
        public String getTelephone() {
            return telephone;
        }
        public void setTelephone(String telephone) {
            this.telephone = telephone;
        }
        public String getEmail() {
            return email;
        }
        public void setEmail(String email) {
            this.email = email;
        }
        public String getJob() {
            return job;
        }
        public void setJob(String job) {
            this.job = job;
        }
        private String telephone;
        private String email;
        private String job;
}
```

创建接口 UserRepository，代码如例 12-3 所示。

【例 12-3】 接口 UserRepository 的代码。

```java
package com.bookcode.backendforreactjs;
import org.springframework.data.jpa.repository.JpaRepository;
public interface UserRepository extends JpaRepository<User,Integer> {
}
```

创建类 UserController，代码如例 12-4 所示。

【例 12-4】 类 UserController 的代码。

```java
package com.bookcode.backendforreactjs;
import org.springframework.web.bind.annotation.*;
import javax.annotation.Resource;
import java.util.List;
//注意跨域问题
@CrossOrigin(origins = {"http://localhost:8080", "null"})
@RestController
@RequestMapping("/api/users")
public class UserController {
    @Resource
    private UserRepository userRepository;
    @GetMapping//查询
    public List<User> getList() {
        return userRepository.findAll();
    }
    @PostMapping//修改和增加
    public User addUser(@RequestBody User user) {
        return userRepository.save(user);
    }
    @DeleteMapping(value = "/{uid}")//删除
    public void delUser(@PathVariable("uid") Integer uid) {
        userRepository.deleteById(uid);
    }
}
```

12.2.3 修改后端配置文件

修改配置文件 application.properties，代码如例 12-5 所示。注意，在本例中使用的 MySQL 版本是 8.x，若使用其他版本的 MySQL（如 5.x），例 12-5 所示中的代码需要调整。

【例 12-5】 修改后的配置文件 application.properties 的代码。

```
spring.datasource.driver-class-name=com.mysql.cj.jdbc.Driver
spring.datasource.url=jdbc:mysql://localhost:3306/db_mediablog?serverTimezone=GMT&useUnicode=true&characterEncoding=UTF-8&useSSL=false
spring.jpa.hibernate.ddl-auto=none
spring.datasource.username=root
spring.datasource.password=ws780125
spring.jpa.show-sql=true
```

12.2.4 数据库文件 db_mediablog.sql

创建完数据库 db_mediablog 之后，按照数据库文件 db_mediablog.sql 创建表 user 并插入记录数据，代码如例 12-6 所示。

【例 12-6】 数据库文件 db_mediablog.sql 的代码。

```
DROP TABLE IF EXISTS 'user';
CREATE TABLE 'user' (
  'uid' int(11) NOT NULL AUTO_INCREMENT,
  'name' varchar(20) DEFAULT NULL,
  'email' varchar(30) DEFAULT NULL,
  'telephone' varchar(30) DEFAULT NULL,
  'job' varchar(30) DEFAULT NULL,
  'age' varchar(30) DEFAULT NULL,
  PRIMARY KEY ('uid')
) ENGINE=InnoDB AUTO_INCREMENT=3 DEFAULT CHARSET=utf8;
INSERT INTO 'user' VALUES ('1', '张三丰', 'z@sohu.com', '12345678901', '老师', '23');
INSERT INTO 'user' VALUES ('2', '李斯', 'lisi@163.com', '12345678901', '程序员', '32');
```

12.2.5 运行后端 Spring Boot 程序

运行入口类 BackendforreactjsApplication，成功启动自带的内置 Tomcat。在浏览器中输入 localhost:8080/api/users 后，效果如图 12-1 所示。

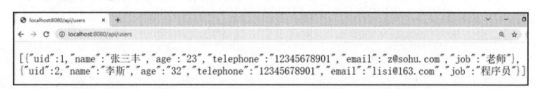

图 12-1 在浏览器中输入 localhost:8080/api/users 后的运行效果

12.3 React 作为前端的开发

12.3.1 修改文件 index.js

继续在第 11 章项目 reactjsbook 的基础上进行开发，修改项目中 src 目录下的文件 index.js 代码，代码如例 12-7 所示。

【例 12-7】 src 目录下文件 index.js 被修改后的代码。

```
import React from 'react';
import ReactDOM from 'react-dom';
import './index.css';
import HelloWorld from './components/HelloWorld';
import reportWebVitals from './reportWebVitals';
ReactDOM.createRoot(document.getElementById('root')).render(
  <React.StrictMode>
    <HelloWorld />
  </React.StrictMode>
);
```

12.3.2 创建组件

在 reactjsbook\src\components 目录下创建文件 HelloWorld.js，代码如例 12-8 所示。

【例 12-8】 文件 HelloWorld.js 的代码。

```
import ApplicationIndex from "./ApplicationIndex";
function HelloWorld() {
    return (
        <div>
            <ApplicationIndex/>
        </div>
    );
}
export default HelloWorld;
```

在 reactjsbook\src\components 目录下创建文件 ApplicationIndex.js，代码如例 12-9 所示。

【例 12-9】 文件 ApplicationIndex.js 的代码。

```
import UserList from "./UserList";
function ApplicationIndex() {
    return (
        <div>
            <UserList/>
        </div>
    );
}
export default ApplicationIndex;
```

在 reactjsbook\中 src\components 目录下创建文件 UserList.js，代码如例 12-10 所示。

【例 12-10】 文件 UserList.js 的代码。

```
import React, {Component} from 'react';
import 'isomorphic-fetch';              //isomorphic-fetch 需要用 npm 安装
import {Button} from 'react-bootstrap';  //react-bootstrap 需要用 npm 安装
export default class userList extends Component {
    constructor() {
        super();
        this.state = {}
    }
    async componentDidMount() {
        let users = await (await fetch('/api/users')).json();//从 Spring Boot
后端获取 JSON 数据
        this.setState({users});
    }
    render() {
        let {users = []} = this.state;
        return (
            <div>
                <table className='table' border="1">
```

```
                <thead>
                <tr>
                    <th>id</th>
                    <th>姓名</th>
                    <th>年龄</th>
                    <th>电话</th>
                    <th>邮箱</th>
                    <th>职位</th>
                    <th>编辑</th>
                </tr>
                </thead>
                <tbody>
                {users.map((({uid, name, age, telephone, job, email}) =>
                    <tr key={uid}>
                        <td>{uid}</td>
                        <td>{name}</td>
                        <td>{age}</td>
                        <td>{telephone}</td>
                        <td>{email}</td>
                        <td>{job}</td>
                        <td><Button onClick = {() => {
                            this.setState({users});
                            alert("welcome")
                        }}>配置</Button></td>
                    </tr>
                ))}
                </tbody>
            </table>
        </div>
    );
  }
}
```

12.3.3 修改前端配置文件

修改项目 reactjsbook 中根目录下文件 package.json 的代码如例 12-11 所示。在此代码中主要增加了一条代理（proxy）信息，如例 12-11 中倒数第 2 行代码所示。代码中包的版本信息请参考本书附的原代码。

【例 12-11】 修改后的文件 package.json 的代码。

```
{
  "name": "reactjsbook",
  "version": "0.1.0",
  "private": true,
  "dependencies": {
    "@testing-library/jest-dom": "^5.16.4",
    "@testing-library/react": "^13.3.0",
    "@testing-library/user-event": "^13.5.0",
```

```
    "isomorphic-fetch": "^3.0.0",
    "react": "^18.2.0",
    "react-bootstrap": "^2.4.0",
    "react-dom": "^18.2.0",
    "react-scripts": "5.0.1",
    "web-vitals": "^2.1.4"
  },
  "scripts": {
    "start": "react-scripts start",
    "build": "react-scripts build",
    "test": "react-scripts test",
    "eject": "react-scripts eject"
  },
  "eslintConfig": {
    "extends": [
      "react-app",
      "react-app/jest"
    ]
  },
  "browserslist": {
    "production": [
      ">0.2%",
      "not dead",
      "not op_mini all"
    ],
    "development": [
      "last 1 chrome version",
      "last 1 firefox version",
      "last 1 safari version"
    ]
  },
  "proxy": "http://localhost:8080"
}
```

在此开发过程中若缺少某个包，则可以使用 NPM 命令安装对应的包。例如，为了安装 isomorphic-fetch 包，可以执行如例 12-12 所示的命令。也可以用 Yarn 命令来安装对应的包，具体方法请参考 10.1 节的内容。

【例 12-12】 安装 isomorphic-fetch 包的命令。

```
npm install isomorphic-fetch
```

12.3.4 运行前端 React 程序

运行项目 reactjsbook，在浏览器中输入 127.0.0.1:3000，效果如图 12-2 所示。

与此同时，MySQL 数据库可视化工具 Navicat for MySQL 中显示数据库中 user 表的数据如图 12-3 所示。对比图 12-1～图 12-3 可以发现，三者的数据是一致的。

图 12-2　在浏览器中输入 127.0.0.1:3000 后的运行效果

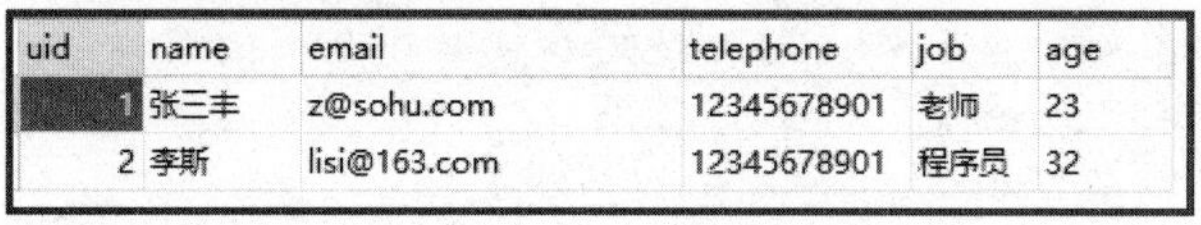

图 12-3　工具 Navicat for MySQL 中显示数据库中 user 表数据的效果

习题 12

一、简答题

1. 简述对 Spring 的理解。
2. 简述对 Spring Boot 特点的理解。

二、实验题

1. 完成 Spring Boot 作为后端的开发。
2. 完成 React 作为前端的开发。

第13章

React与Python框架的整合开发

本章先介绍 React 与 Django 默认数据库 dbsqlite3 整合开发的示例,此示例演示了 React 用 axios 方式访问 Django 的 API;再介绍 React 与 Flask 的整合开发示例,此示例演示了将 React 程序代码打包后放入 Flask 项目中进行应用开发的过程。

13.1 React 与 Django 的整合开发

视频讲解

13.1.1 Django 作为后端开发

参考本书电子版附录 A.3 节或其他资源,安装 Python 和 PyCharm,再用 PyCharm 创建 Django 项目 backend,并用如例 13-1 所示的命令在项目 backend 中创建应用 App。

【例 13-1】 在项目 backend 中创建应用 App 的命令。

```
python manage.py startapp App
```

修改项目 backend 中的 backend 目录下文件 settings.py 代码,再去掉注释,如例 13-2 所示。

【例 13-2】 backend 目录下文件 settings.py 被修改后的代码。

```
import os
BASE_DIR = os.path.dirname(os.path.dirname(os.path.abspath(__file__)))
SECRET_KEY = 'g%#f1qbvr1*t#2oz4)d8+m7*=x(6)2^^rp6y5jj_e5fsifao#j'
DEBUG = True
ALLOWED_HOSTS = []
INSTALLED_APPS = [
    'django.contrib.admin',
    'django.contrib.auth',
    'django.contrib.contenttypes',
```

```python
        'django.contrib.sessions',
        'django.contrib.messages',
        'django.contrib.staticfiles',
        'rest_framework',       //新增
        'corsheaders',          //新增
        'App',                  //新增
]
MIDDLEWARE = [
    'django.middleware.security.SecurityMiddleware',
    'django.contrib.sessions.middleware.SessionMiddleware',
    'django.middleware.common.CommonMiddleware',
    'django.middleware.csrf.CsrfViewMiddleware',
    'django.contrib.auth.middleware.AuthenticationMiddleware',
    'django.contrib.messages.middleware.MessageMiddleware',
    'django.middleware.clickjacking.XFrameOptionsMiddleware',
    'corsheaders.middleware.CorsMiddleware'  //新增
]
//新增跨域资源共享(Cross-Origin Resource Sharing)
CORS_ORIGIN_ALLOW_ALL = False
CORS_ORIGIN_WHITELIST = (
    ['http://127.0.0.1:*']
)
REST_FRAMEWORK = {
    'DEFAULT_AUTHENTICATION_CLASSES': [],
    'DEFAULT_PERMISSION_CLASSES': [],
}
ROOT_URLCONF = 'backend.urls'
TEMPLATES = [
    {
        'BACKEND': 'django.template.backends.django.DjangoTemplates',
        'DIRS': [],
        'APP_DIRS': True,
        'OPTIONS': {
            'context_processors': [
                'django.template.context_processors.debug',
                'django.template.context_processors.request',
                'django.contrib.auth.context_processors.auth',
                'django.contrib.messages.context_processors.messages',
            ],
        },
    },
]
WSGI_APPLICATION = 'backend.wsgi.application'
DATABASES = {
    'default': {
        'ENGINE': 'django.db.backends.sqlite3',
        'NAME': os.path.join(BASE_DIR, 'db.sqlite3'),
    }
}
AUTH_PASSWORD_VALIDATORS = [
```

```
    {
        'NAME': 'django.contrib.auth.password_validation.
UserAttributeSimilarityValidator',
    },
    {
        'NAME': 'django.contrib.auth.password_validation.
MinimumLengthValidator',
    },
    {
        'NAME': 'django.contrib.auth.password_validation.
CommonPasswordValidator',
    },
    {
        'NAME': 'django.contrib.auth.password_validation.
NumericPasswordValidator',
    },
]
LANGUAGE_CODE = 'en-us'
TIME_ZONE = 'UTC'
USE_I18N = True
USE_L10N = True
USE_TZ = True
STATIC_URL = '/static/'
DEFAULT_AUTO_FIELD = 'django.db.models.BigAutoField'
```

修改项目 backend 中的 backend 目录下文件 urls.py，代码如例 13-3 所示。

【例 13-3】 backend 目录下文件 urls.py 被修改后的代码。

```
from django.contrib import admin
from django.urls import path, include
from rest_framework import routers
from App import views
router = routers.DefaultRouter()
router.register('Testing', views.TestingView, 'Testing')
urlpatterns = [
    path('admin/', admin.site.urls),
    path('api/', include(router.urls))  //新增
]
```

修改项目 backend 中的 App 目录下文件 apps.py，代码如例 13-4 所示。

【例 13-4】 App 目录下文件 apps.py 被修改后的代码。

```
from django.apps import AppConfig
class AppConfig(AppConfig):
    default_auto_field = 'django.db.models.BigAutoField'
    name = 'App'
```

修改项目 backend 中的 App 目录下文件 models.py，代码如例 13-5 所示。

【例 13-5】 App 目录下文件 models.py 被修改后的代码。

```python
from django.db import models
class Testing(models.Model):
    text = models.TextField()
    def __str__(self):
        return self.text
```

修改项目 backend 中的 App 目录下文件 serializers.py，代码如例 13-6 所示。

【例 13-6】 App 目录下文件 serializers.py 被修改后的代码。

```python
from rest_framework import serializers
from .models import Testing
class TestingSerializer(serializers.ModelSerializer):
    class Meta:
        model = Testing
        fields = '__all__'
```

修改项目 backend 中的 App 目录下文件 views.py，代码如例 13-7 所示。

【例 13-7】 App 目录下文件 views.py 被修改后的代码。

```python
from rest_framework import viewsets
from .serializers import TestingSerializer
from .models import Testing
class TestingView(viewsets.ModelViewSet):
    serializer_class = TestingSerializer
    queryset = Testing.objects.all()
```

用如例 13-8 所示的命令在项目 backend 中自动生成数据库。

【例 13-8】 在项目 backend 中自动生成数据库的命令。

```
python manage.py makemigrations
python manage.py migrate
```

13.1.2　运行后端 Django 程序

运行项目 backend，在浏览器中输入 127.0.0.1:8000/api/，效果如图 13-1 所示。

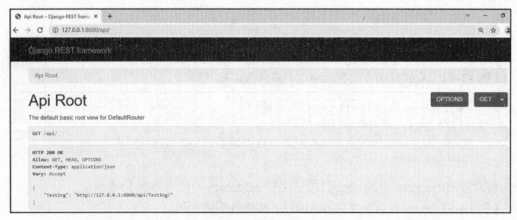

图 13-1　运行项目 backend 后在浏览器中输入 127.0.0.1:8000/api/ 的效果

13.1.3　React 作为前端开发

参考 8.2 节用 WebStorm 创建 React 项目 frontend，修改项目 frontend 中的 src 目录下文件 App.js，代码如例 13-9 所示。

【例 13-9】　src 目录下文件 App.js 被修改后的代码。

```javascript
import React, { Component } from "react";
import "./App.css";
import TextItem from "./TextItem";
import axios from "axios"; //安装包 axios
axios.defaults.xsrfCookieName = "csrftoken";
axios.defaults.xsrfHeaderName = "X-CSRFToken";
class App extends Component {
  state = {
    value: "",
    textList: []
  };
  componentDidMount() {
    this._renderText();
  }
  render() {
    const { textList } = this.state;
    console.log(textList);
    return (
      <div className="App">
        <h1>待办事项小工具</h1>
        <div>
          <label>
            输入新任务（待办事项）
            <input
              type="text"
              value={this.state.value}
              onChange={this._handleChange}
            />
          </label>
          <button onClick={this._handleSubmit}>确定</button>
        </div>
        <h2>尚未处理事项</h2>
        {textList.map((text, index) => {
          return (
            <TextItem
              text={text.text}
              key={index}
              id={text.id}
              handleClick={this._deleteText}
            />
          );
```

```
        }))}
      </div>
    );
  }
  _handleChange = event => {
    this.setState({ value: event.target.value });
  };
  _handleSubmit = () => {
    const { value } = this.state;
    axios
      .post("/api/Testing/", { text: value })
      .then(res => this._renderText());
  };
  _renderText = () => {
    axios
      .get("/api/Testing/")
      .then(res => this.setState({ textList: res.data }))
      .catch(err => console.log(err));
  };
  _deleteText = id => {
    axios.delete(`/api/Testing/${id}`).then(res => this._renderText());
  };
}
export default App;
```

在 frontend\src 目录下创建文件 TextItem.js，代码如例 13-10 所示。

【例 13-10】 文件 TextItem.js 的代码。

```
import React from "react";
const TextItem = ({ text, id, handleClick }) => {
  return (
    <div>
      <p>{text}</p>
      <button onClick={() => handleClick(id)}>删除办结事项</button>
    </div>
  );
};
export default TextItem;
```

修改项目 frontend 中根目录下文件 package.json，代码如例 13-11 所示。主要是增加了一条代理（proxy）信息，如例 13-11 中第 4 行代码所示。代码中包的版本可参考本书附的源代码。

【例 13-11】 根目录下文件 package.json 被修改后的代码。

```
{
  "name": "frontend",
  "version": "0.1.0",
  "private": true,
  "proxy": "http://localhost:8000",
```

```
    "dependencies": {
      "@testing-library/jest-dom": "^5.16.4",
      "@testing-library/react": "^13.3.0",
      "@testing-library/user-event": "^13.5.0",
      "axios": "^0.27.2",
      "bootstrap": "^5.1.3",
      "less": "^4.1.2",
      "lodash": "^4.17.21",
      "react": "^18.2.0",
      "react-axios": "^2.0.5",
      "react-dom": "^18.2.0",
      "react-router-dom": "^6.0.2",
      "react-scripts": "5.0.1",
      "reactstrap": "^9.0.1",
      "web-vitals": "^2.1.4"
    },
    "scripts": {
      "start": "react-scripts start",
      "build": "react-scripts build",
      "test": "react-scripts test",
      "eject": "react-scripts eject"
    },
    "eslintConfig": {
      "extends": [
        "react-app",
        "react-app/jest"
      ]
    },
    "browserslist": {
      "production": [
        ">0.2%",
        "not dead",
        "not op_mini all"
      ],
      "development": [
        "last 1 chrome version",
        "last 1 firefox version",
        "last 1 safari version"
      ]
    }
}
```

在此开发过程中若缺少某个包，则可以使用 npm 命令安装对应的包。例如，为了安装 axios 包，可以执行如例 13-12 所示的命令。axios 是一个基于 promise 的 HTTP 包，它可以运行在浏览器和 Node.js 环境中，利用它可以很容易地发生 HTTP 请求。

【例 13-12】 安装 axios 包的命令。

```
npm install axios
```

13.1.4 运行前端 React 程序

运行项目 frontend，在浏览器中输入 localhost:3000，效果如图 13-2 所示。在图 13-2 所示中的文本框中输入"1.需求分析"，效果如图 13-3 所示。单击图 13-3 所示中的"确定"按钮后的效果如图 13-4 所示。在图 13-4 所示中的文本框中输入"2.总体设计"并单击"确定"按钮后的效果如图 13-5 所示。

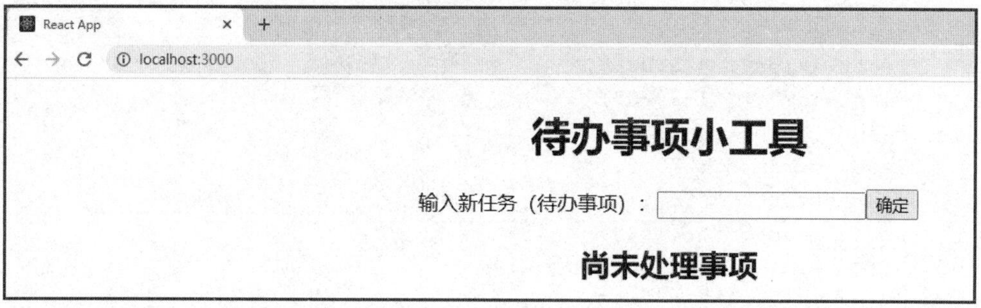

图 13-2　运行项目 frontend 后在浏览器中输入 localhost:3000 的效果

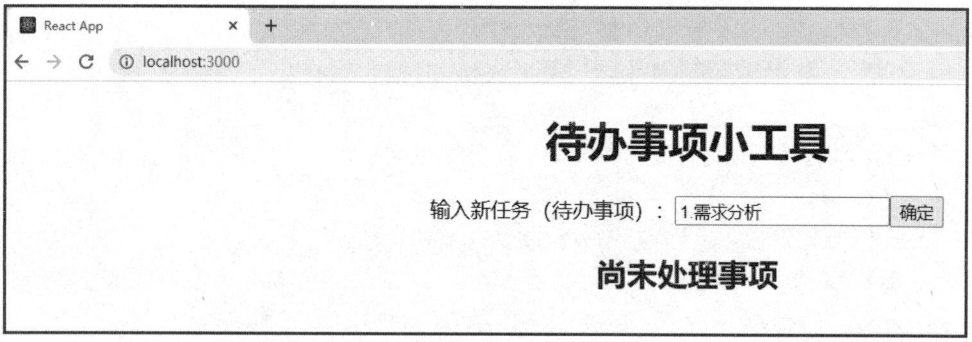

图 13-3　在图 13-2 所示中的文本框中输入"1.需求分析"的效果

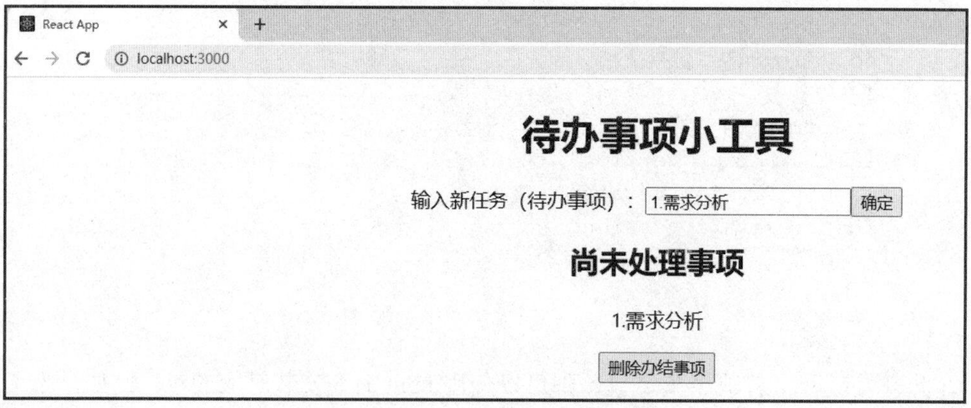

图 13-4　单击图 13-3 所示中的"确定"按钮后的效果

这些在 React 前端的操作，会写入 Django 后端的数据库中。本例采用的数据库是 Django 默认 db.sqlite3 数据库，通过可视化工具 SQLite Expert Personal 执行如例 13-13 所示的 SQL 语句，可以看到刚输入的两条任务（待办事项）被成功地写入后端数据库中，效果如图 13-6 所示。

图 13-5　在图 13-4 所示中的文本框中输入"2.总体设计"并单击"确定"按钮后的效果

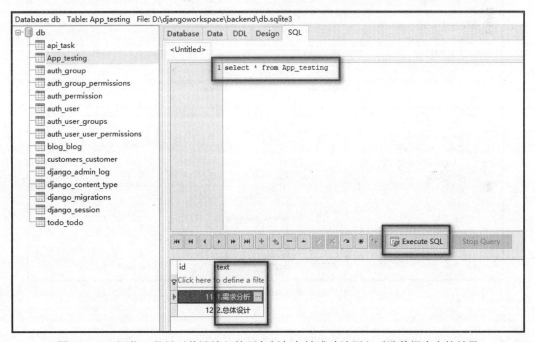

图 13-6　可视化工具显示前端输入的两条新任务被成功地写入后端数据库中的效果

【例 13-13】　获取数据库内容的 SQL 语句。

```
select * from App_testing
```

单击图 13-5 所示中"1.需求分析"下方的"删除办结事项"按钮，可删除此项任务，效果如图 13-7 所示。此时，通过可视化工具 SQLite Expert Personal 执行如例 13-13 的 SQL 语句，可以看到刚删除的任务已经在后端数据库中也已经删除，效果如图 13-8 所示。

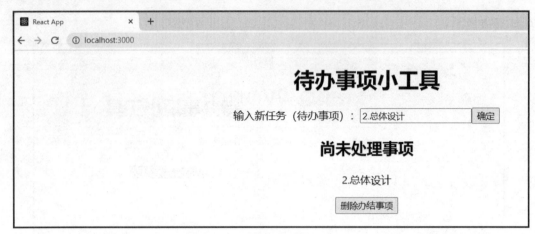

图 13-7　单击图 13-5 所示中 "1.需求分析" 下方的 "删除办结事项" 按钮后的效果

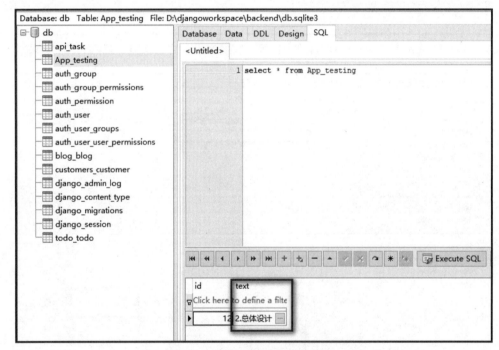

图 13-8　数据库删除任务的效果

视频讲解

13.2　React 与 Flask 的整合开发

13.2.1　Flask 作为后端开发和运行后端 Flask 程序

参考本书电子版附录 A.3 或其他资源，用 PyCharm 创建 Flask 项目 flaskreact，在运行该项目后，在浏览器中输入 localhost:5000，效果如图 13-9 所示。使用 PyCharm 创建 Flask 项目时，搭建环境并创建 HelloWorld 应用较为简单。

图 13-9　用 PyCharm 创建 Flask 的 HelloWorld 应用的效果

13.2.2　React 作为前端开发

参考 8.2 节用 WebStorm 创建项目 reactforflask，修改项目 src 目录下的文件 index.js，代码如例 13-14 所示。

【例 13-14】　src 目录下文件 index.js 被修改后的代码。

```
import React from 'react';
import ReactDOM from 'react-dom';
import './index.css';
import reportWebVitals from './reportWebVitals';
import HelloWorld from './HelloWorld';
ReactDOM.createRoot(document.getElementById('root')).render(
  <React.StrictMode>
    <HelloWorld />
  </React.StrictMode>
);
```

在 reactforflask\src 目录下创建文件 HelloWorld.js，代码如例 13-15 所示。

【例 13-15】　文件 HelloWorld.js 的代码。

```
function HelloWorld () {
    return (
        <div className="container">
            <h3 className="center-align">
                下面两行文字自 React 前端程序（实现基础组件）：<br/>
                <span className="waves-effect waves-light btn">
                    <i className="material-icons right">斜体部分</i><br/>
                    正体部分
                    <hr/>
                </span>
            </h3>
        </div>
    );
}
export default HelloWorld;
```

13.2.3　运行前端 React 程序

运行项目 reactforflask，在浏览器中输入 localhost:3000，效果如图 13-10 所示。

图 13-10　运行项目 reactforflask 后在浏览器中输入 localhost:3000 的效果

13.2.4　打包前端 React 程序代码并手工复制到后端

使用 npm 命令可以打包 React 程序代码，如例 13-16 所示的打包前端 React 程序的命令。

【例 13-16】 打包前端 React 程序的命令。

```
npm run build
```

把前端 React 程序代码打包后，将 build 目录中的入口文件 index.html 复制到 Flask 后端程序的 template 目录（创始创建项目时该目录为空目录）下。把 build 目录中其他的 js、css、图片等文件复制到 Flask 后端程序的 static 目录（开始创建项目时该目录为空目录）下。复制后的效果如图 13-11 所示。参考源代码或视频修改文件 app.py。

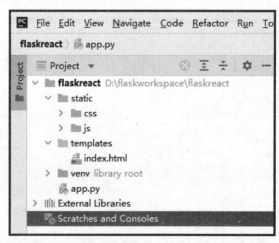

图 13-11　打包前端 React 程序并复制到后端 flaskreact 后目录文件结构的效果

13.2.5　运行后端 Flask 程序

运行该项目后在浏览器中输入 localhost:5000，效果如图 13-12 所示。对比图 13-9、图 13-10 和图 13-12 可以发现，图 13-10（端口号为 3000）和图 13-12（端口号为 5000）的输出文字相同而端口不同；图 13-9（端口号为 5000）和图 13-12（端口号为 5000）的端口相同而输出文字不同。通过这两项的对比均可以说明前后端的整合是成功的。

图 13-12　运行后端项目后在浏览器中输入 localhost:5000 的效果

习题 13

一、实验题

1. 完成 React 与 Django 的整合开发。
2. 完成 React 与 Flask 的整合开发。

二、简答题

在完成 11.3 节和 13.1 节应用开发的基础上，对比分析两种实现方式的差别，加深对前后端分离的理解。

第 14 章

React 与 Go 的整合开发

视频讲解

本章介绍 React 与 Go、MySQL 数据库的整合开发，实现一个简易个人信息查看功能的示例。本示例演示了使用 XMLHttpRequest 对象与服务器交互的情况。通过 XMLHttpRequest 可以在不刷新页面的情况下，请求特定 URL 获取数据。通过使用 XMLHttpRequest 的工作方式，允许网页在不影响用户操作的情况下，更新页面的局部内容。

14.1 Go 作为后端的开发

14.1.1 创建项目 server 和入口文件

参考本书电子版附录 A.4 或其他资源，先安装 Go 和 GoLand，再用 GoLand 创建项目 server，并在项目 server 根目录下创建文件 main.go，代码如例 14-1 所示。

【例 14-1】 文件 main.go 的代码。

```
package main
import (
    "github.com/gorilla/handlers"
    "github.com/gorilla/mux"
    "github.com/pkg/errors"
    "github.com/sirupsen/logrus"
    "net/http"
    "sever/api"
    "sever/db"
)
const (
    portAPI = "8080"
)
func main() {
```

```go
        dbMySQL := createDBConnection()
        defer dbMySQL.Close()
        startAPIServer(dbMySQL)
    }
    func createDBConnection() db.DB {
        mySQLPass :="ws780125"
        if mySQLPass == "" {
            logrus.Fatal("Env var MYSQL_PASS was not set!")
        }
        dbMySQL, err := db.NewMySqlDB(mySQLPass)
        if err != nil {
           logrus.Fatal(errors.Wrapf(err, "failed to start MySQL"))
        }
        return dbMySQL
    }
    func startAPIServer(dbMySQL db.DB) {
        appAPI := api.NewDefaultAPI(dbMySQL)
        router := mux.NewRouter()
        router.HandleFunc("/app/people", appAPI.ListPersons).Methods(http.MethodGet)
        router.HandleFunc("/app/people", appAPI.CreatePerson).Methods(http.MethodPost)
        router.HandleFunc("/app/people", appAPI.DeletePerson).Methods(http.MethodDelete)
        router.HandleFunc("/app/people/{email}", appAPI.GetPerson).Methods(http.MethodGet)
        router.HandleFunc("/app/people/{email}", appAPI.UpdatePerson).Methods(http.MethodPut)
        logrus.Info("REST API server listening on port " + portAPI)
        if err := http.ListenAndServe(":8080", handlers.CORS(handlers.AllowedHeaders([]string{"X-Requested-With", "Content-Type", "Authorization"}), handlers.AllowedMethods([]string{"GET", "POST", "PUT", "DELETE", "HEAD", "OPTIONS"}), handlers.AllowedOrigins([]string{"*"}))(router)); err != nil {
            logrus.Fatal("Failed to listen and serve on port " + portAPI)
        }
    }
```

14.1.2 创建 API

在项目 server 中创建 api 包，在 api 包下创建文件 api.go，代码如例 14-2 所示。

【例 14-2】文件 api.go 的代码示例。

```go
package api
import (
    "encoding/json"
    "net/http"
    "path"
    "sever/db"
```

```go
    )
    type API interface {
        CreatePerson(w http.ResponseWriter, r *http.Request)
        DeletePerson(w http.ResponseWriter, r *http.Request)
        ListPersons(w http.ResponseWriter, r *http.Request)
        GetPerson(w http.ResponseWriter, r *http.Request)
        UpdatePerson(w http.ResponseWriter, r *http.Request)
    }
    // DefaultAPI default implementation of API.
    type DefaultAPI struct {
        db db.DB
    }
    // NewDefaultAPI creates a new DefaultAPI.
    func NewDefaultAPI(db db.DB) *DefaultAPI {
        return &DefaultAPI{db: db}
    }
    func (dAPI *DefaultAPI) CreatePerson(w http.ResponseWriter, r *http.Request) {
        w.Header().Set("Content-Type", "application/json")
        person, err := dAPI.mapPersonPayload(r)
        if err != nil {
            w.WriteHeader(http.StatusBadRequest)
            resp, _ := json.Marshal(Response{Error: "Invalid payload."})
            w.Write(resp)
            return
        }
        if err := dAPI.db.CreatePerson(person); err != nil {
            w.WriteHeader(http.StatusInternalServerError)
            resp, _ := json.Marshal(Response{Error: "There was a problem with the
server."})
            w.Write(resp)
            return
        }
        w.WriteHeader(http.StatusOK)
        resp, _ := json.Marshal(Response{Message: "Successfully created the
person."})
        w.Write(resp)
    }
    func (dAPI *DefaultAPI) DeletePerson(w http.ResponseWriter, r *http.Request) {
        w.Header().Set("Content-Type", "application/json")
        if err := r.ParseForm(); err != nil {
            w.WriteHeader(http.StatusInternalServerError)
            resp, _ := json.Marshal(Response{Error: err.Error()})
            w.Write(resp)
            return
        }
        email := r.FormValue("email")
        if email == "" {
            w.WriteHeader(http.StatusBadRequest)
            resp, _ := json.Marshal(Response{Error: "Missing parameter 'email'."})
```

```go
            w.Write(resp)
            return
        }
        if err := dAPI.db.DeletePerson(email); err != nil {
            w.WriteHeader(http.StatusInternalServerError)
            resp, _ := json.Marshal(Response{Error: "There was a problem with the server."})
            w.Write(resp)
            return
        }
        w.WriteHeader(http.StatusOK)
        resp, _ := json.Marshal(Response{Message: "Successfully deleted the person."})
        w.Write(resp)
    }
    func (dAPI *DefaultAPI) ListPersons(w http.ResponseWriter, r *http.Request) {
        w.Header().Set("Content-Type", "application/json")
        if err := r.ParseForm(); err != nil {
            w.WriteHeader(http.StatusInternalServerError)
            resp, _ := json.Marshal(Response{Error: err.Error()})
            w.Write(resp)
            return
        }
        orderBy := r.FormValue("orderBy")
        if orderBy == "" {
            orderBy = "email"
        } else if orderBy != "name" && orderBy != "email" {
            w.WriteHeader(http.StatusBadRequest)
            resp, _ := json.Marshal(Response{Error: "Unsupported sorting column: " + orderBy})
            w.Write(resp)
            return
        }
        persons, err := dAPI.db.ListPersons(orderBy)
        if err != nil {
            w.WriteHeader(http.StatusInternalServerError)
            resp, _ := json.Marshal(Response{Error: err.Error()})
            w.Write(resp)
            return
        }
        w.WriteHeader(http.StatusOK)
        resp, _ := json.Marshal(persons)
        w.Write(resp)
    }
    func (dAPI *DefaultAPI) GetPerson(w http.ResponseWriter, r *http.Request) {
        w.Header().Set("Content-Type", "application/json")
        email := path.Base(r.RequestURI)
        person, err := dAPI.db.RetrievePerson(email)
        if err != nil {
```

```go
            w.WriteHeader(http.StatusInternalServerError)
            resp, _ := json.Marshal(Response{Error: err.Error()})
            w.Write(resp)
            return
        }
        w.WriteHeader(http.StatusOK)
        resp, _ := json.Marshal(person)
        w.Write(resp)
    }
    func (dAPI *DefaultAPI) UpdatePerson(w http.ResponseWriter, r *http.Request){
        w.Header().Set("Content-Type", "application/json")
        email := path.Base(r.RequestURI)
        updatedPerson, err := dAPI.mapPersonPayload(r)
        if err != nil {
            w.WriteHeader(http.StatusBadRequest)
            resp, _ := json.Marshal(Response{Error: "Invalid payload."})
            w.Write(resp)
            return
        }
        if err := dAPI.db.UpdatePerson(email, updatedPerson); err != nil {
            w.WriteHeader(http.StatusInternalServerError)
            resp, _ := json.Marshal(Response{Error: "There was a problem with the server."})
            w.Write(resp)
            return
        }
        w.WriteHeader(http.StatusOK)
        resp, _ := json.Marshal(Response{Message: "Successfully updated the person."})
        w.Write(resp)
    }
```

14.1.3 创建工具类

在 api 包下创建文件 util.go，代码如例 14-3 所示。

【例 14-3】 文件 util.go 的代码。

```go
package api
import (
    "encoding/json"
    "errors"
    "io/ioutil"
    "net/http"
    "sever/db"
)
type Response struct {
    Message string `json:"message,omitempty"`
    Error   string `json:"error,omitempty"`
```

```go
}
// mapPersonPayload()根据HTTP请求返回person
func (dAPI *DefaultAPI) mapPersonPayload(r *http.Request) (*db.Person, error) {
    body, err := ioutil.ReadAll(r.Body)
    if err != nil {
        return nil, err
    }
    person := db.Person{}
    if err := json.Unmarshal(body, &person); err != nil {
        return nil, err
    }
    return &person, dAPI.validatePerson(&person)
}
// validatePerson()方法检验person数据
func (dAPI *DefaultAPI) validatePerson(person *db.Person) error {
    if person.Email == "" {
        return errors.New("Missing mandatory field 'email'")
    }
    return nil
}
```

14.1.4 创建数据库处理类

在项目 server 中创建 db 包，在 db 包下创建文件 db.go，代码如例 14-4 所示。

【例 14-4】 文件 db.go 的代码。

```go
package db
import (
    "database/sql"
    _ "github.com/go-sql-driver/mysql"
    "github.com/sirupsen/logrus"
    "time"
)
type DB interface {
    CreatePerson(person *Person) error
    DeletePerson(email string) error
    ListPersons(orderBy string) ([]*Person, error)
    RetrievePerson(email string) (*Person, error)
    UpdatePerson(email string, updatedPerson *Person) error
    Close()
}
type MySqlDB struct {
    db *sql.DB
}
func NewMySqlDB(pwd string) (*MySqlDB, error) {
    db, err := TryConnect("root:"+pwd+"@tcp(localhost:3306)/godatabase", 3, 5)
    if err != nil {
```

```go
            return nil, err
        }
        return &MySqlDB{db: db}, nil
    }
    // 连接数据库
    func TryConnect(dsn string, delay, retries int) (*sql.DB, error) {
        db, err := sql.Open("mysql", dsn)
        for ; err != nil && retries > 0; retries-- {
            time.Sleep(time.Second * time.Duration(delay))
            db, err = sql.Open("mysql", dsn)
        }
        return db, err
    }
    // 关闭数据库连接
    func (my *MySqlDB) Close() {
        if err := my.db.Close(); err != nil {
            logrus.Panic("Failed to close MySQL db.")
        }
    }
    func (my *MySqlDB) CreatePerson(person *Person) error {
        insertQuery, err := my.db.Prepare("INSERT INTO Persons VALUES(?,?,?,?,?);")
        if err != nil {
            return err
        }
        if _, err := insertQuery.Exec(person.Name, person.Age, person.Balance, person.Email, person.Address); err != nil {
            return err
        }
        return nil
    }
    func (my *MySqlDB) DeletePerson(email string) error {
        deleteQuery, err := my.db.Prepare("DELETE FROM Persons WHERE Email = ?;")
        if err != nil {
            return err
        }
        if _, err = deleteQuery.Exec(email); err != nil {
            return err
        }
        return nil
    }
    func (my *MySqlDB) ListPersons(orderBy string) ([]*Person, error) {
        rows, err := my.db.Query("SELECT * FROM Persons ORDER BY " + orderBy + " DESC;")
        if err != nil {
            return nil, err
        }
        persons := []*Person{}
        for rows.Next() {
            person := &Person{}
```

```go
        if err := rows.Scan(&person.Name, &person.Age, &person.Balance,
&person.Email, &person.Address); err != nil {
            return nil, err
        }
        persons = append(persons, person)
    }
    return persons, nil
}
func (my *MySqlDB) RetrievePerson(email string) (*Person, error) {
    selectQuery, err := my.db.Prepare("SELECT * FROM Persons WHERE Email = ?;")
    if err != nil {
        return nil, err
    }
    row := selectQuery.QueryRow(email)
    person := &Person{}
    return person, row.Scan(&person.Name, &person.Age, &person.Balance,
&person.Email, &person.Address)
}
func (my *MySqlDB) UpdatePerson(email string, updatedPerson *Person) error{
    updateQuery, err := my.db.Prepare("UPDATE Persons SET Name = ?, Age = ?,
Balance = ?, Email = ?, Address = ? WHERE Email = ?;")
    if err != nil {
        return err
    }
    if _, err = updateQuery.Exec(updatedPerson.Name, updatedPerson.Age,
updatedPerson.Balance, updatedPerson.Email, updatedPerson.Address, email);
err != nil {
        return err
    }
    return nil
}
```

14.1.5 创建数据类型

在 db 包下创建文件 model.go，代码如例 14-5 所示。

【例 14-5】文件 model.go 的代码。

```go
package db
// Person contains information about a person.
type Person struct {
    Name    string  'json:"name,omitempty"'
    Age     int     'json:"age,omitempty"'
    Balance float32 'json:"balance,omitempty"'
    Email   string  'json:"email"'
    Address string  'json:"address,omitempty"'
}
```

14.1.6 数据库文件 godatabase.sql

创建完数据库 godatabase 之后，按照数据库文件 godatabase.sql 创建表 persons 并插入数值，代码如例 14-6 所示。

【例 14-6】 数据库文件 godatabase.sql 的代码。

```sql
DROP TABLE IF EXISTS 'persons';
CREATE TABLE 'persons' (
  'Name' varchar(255) DEFAULT NULL,
  'Age' int(11) DEFAULT NULL,
  'Balance' double DEFAULT NULL,
  'Email' varchar(255) NOT NULL,
  'Address' varchar(255) DEFAULT NULL,
  PRIMARY KEY ('Email')
) ENGINE=InnoDB DEFAULT CHARSET=utf8mb4 COLLATE=utf8mb4_0900_ai_ci;
INSERT INTO 'persons' VALUES ('李斯', '25', '2500', 'ls@qg.com', '陕西咸阳');
INSERT INTO 'persons' VALUES('王重阳','34','3000.5','wangchongyang@nh.com','山东宁海');
INSERT INTO 'persons' VALUES('王阳明', '46', '5000.5', 'wangyangming@mc.com','贵州龙场');
INSERT INTO 'persons' VALUES('张三丰','28','2000.5','zhangsanfeng@wudang.com','湖北武当');
```

14.1.7 运行后端 Go 程序

运行入口文件 main.go，成功启动 server。在浏览器中输入 localhost:8080/app/people 后，效果如图 14-1 所示。与此同时，MySQL 数据库可视化工具 Navicat for MySQL 显示数据库中 persons 表的数据，效果如图 14-2 所示。对比图 14-1 和图 14-2 可以发现，二者的数据是一致的。

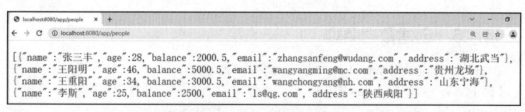

图 14-1　在浏览器中输入 localhost:8080/app/people 后的运行效果

图 14-2　Navicat for MySQL 显示数据库中 persons 表的数据效果

14.2 React 作为前端的开发

14.2.1 创建项目并修改文件 index.js

参考 8.2 节用 WebStorm 创建 React 项目 goreact，修改项目中 src 目录下文件 index.js，代码如例 14-7 所示。

【例 14-7】 src 目录下文件 index.js 被修改后的代码。

```
import React from 'react';
import ReactDOM from 'react-dom';
import reportWebVitals from './reportWebVitals';
import UI from "./UI";
ReactDOM.createRoot(document.getElementById('root')).render(
    <React.StrictMode>
        <UI />
    </React.StrictMode>
);
reportWebVitals();
```

14.2.2 创建用户界面

在 goreact\src 目录下创建文件 UI.js，代码如例 14-8 所示。

【例 14-8】 文件 UI.js 的代码。

```
import React from 'react';
import './UI.css';
import PersonsList from "./PersonsList";
class UI extends React.Component {
  constructor() {
    super()
    this.state = {
      persons: []
    };
    this.getData();
  };
  getData() {
    var xhr = new XMLHttpRequest()
    xhr.addEventListener('load', () => {
      this.setState({
        persons: JSON.parse(xhr.responseText)
      });
    })
    xhr.open('GET', 'http://localhost:8080/app/people?orderBy=email')
    xhr.setRequestHeader("Content-Type", "application/json");
```

```
      xhr.send()
    };
    render() {
      return (
        <div className="UI">
          <header className="UI-header">
            <h1 className="UI-title">个人信息查看</h1>
            <PersonsList persons={this.state.persons} />
          </header>
        </div>
      );
    };
}
export default UI;
```

在 goreact\src 目录下创建文件 UI.css，代码如例 14-9 所示。

【例 14-9】 文件 UI.css 的代码。

```
.UI {
  text-align: center;
}
.UI-header {
  min-height: 100vh;
  display: flex;
  flex-direction: column;
  align-items: center;
  justify-content: center;
  font-size: calc(10px + 2vmin);
  color: white;
}
.UI-title {
  vertical-align: text-top;
  color: #fb616e;
}
```

14.2.3 创建组件

在 goreact\src 目录下创建文件 PersonsList.js，代码如例 14-10 所示。

【例 14-10】 文件 PersonsList.js 的代码。

```
import React from "react";
import Person from "./Person";
function PersonsList(props) {
  return (
    <div>
      {props.persons.map(item =>
        <Person key={item.email} name={item.name} age={item.age} balance=
```

```
{item.balance} email={item.email} address={item.address} checked={item.checked}/>)}
        </div>
    );
}
export default PersonsList;
```

在 goreact\src 目录下创建文件 Person.js，代码如例 14-11 所示。

【例 14-11】 文件 Person.js 的代码。

```
import React from "react";
import "./Person.css";
class Person extends React.Component {
    constructor(props) {
        super(props)
        this.state = {
          checked: localStorage.getItem(props.email)
        };
        this.handleChecked = this.handleChecked.bind(this);
    };
    handleChecked() {
        if (this.getState() === "") {
            this.setState({checked: "checked"});
        } else {
            this.setState({checked: ""});
        }
    };
    getState() {
        return this.state.checked;
    };
    componentDidUpdate(prevProps, prevState) {
        localStorage.setItem(this.props.email, this.state.checked);
    };
    render() {
        return (
            <div className="person">
              <span>{this.props.name}, {this.props.age}, {this.props.balance},
                {this.props.email}, {this.props.address}</span>
    <input type="checkbox" checked={this.state.checked} onChange=
{this.handleChecked}> </input>
            </div>
        )
    };
}
export default Person;
```

在 goreact\src 目录下创建文件 Person.css，代码如例 14-12 所示。

【例 14-12】 文件 Person.css 的代码。

```css
person {
    margin: 10px;
    padding: 10px;
    border: 1px solid #bbb;
    background-color: #eee;
    height:30px;
}
.person span {
    font-size: 1.2em;
    text-decoration: none;
    float: left;
    display: block;
    color: #333;
}
input[type='checkbox'] {
    -webkit-appearance:none;
    width:30px;
    height:30px;
    background:white;
    border-radius:10px;
    border:2px solid #555;
    float: right;
}
input[type='checkbox']:checked {
    background: #036;
}
```

14.2.4 运行前端 React 程序

运行项目 goreact，在浏览器中输入 localhost:3000，效果如图 14-3 所示。对比图 14-1～图 14-3 可以发现，三者的数据是一致的。

图 14-3　在浏览器中输入 localhost:3000 后的运行效果

习题 14

一、实验题

1. 完成 Go 作为后端的开发。
2. 完成 React 作为前端的开发。

二、实验与分析题

结合第 12 章（用 fetch）、第 13 章（用 axios、打包）和本章（用 XMLHttpRequest）前后端整合开发的实现方法，加深对 React 与后端不同整合方法的认识。

第15章

案例——实现一个简易的员工信息管理系统

视频讲解

本章介绍 React 与 Spring Boot、MySQL 数据库整合开发，实现一个简易的员工信息管理系统的案例。本案例还应用了 Material-UI 框架，借助于该框架的应用演示其他框架在 React 应用开发中的应用。Material-UI 是 React 组件，实现了谷歌公司的 Material Design 设计规范。Material Design 是基于传统设计原则开发的一套全新的界面设计语言，包括视觉、运动、互动效果等特性。Material-UI 是一个流行的 React 界面框架，其组件是相互独立的、自支持的，在开发时仅注入当前组件所需要的样式即可。

15.1 Spring Boot 作为后端的开发

15.1.1 创建项目 excase 和实体类

用 IDEA 创建项目 excase，文件 pom.xml 被修改后的代码如例 12-1 所示。

创建类 Employee，代码如例 15-1 所示。

【例 15-1】类 Employee 的代码。

```
package edu.code.model;
import java.time.LocalDate;
import javax.persistence.Column;
import javax.persistence.Entity;
import javax.persistence.GeneratedValue;
import javax.persistence.GenerationType;
import javax.persistence.Id;
import javax.persistence.Table;
@Entity
@Table(name = "tb_emp")
public class Employee {
    @Id
```

```java
    @GeneratedValue(strategy = GenerationType.IDENTITY)
    @Column
    private Integer id;
    @Column
    private String name;
    @Column
    private String department;
    @Column
    private LocalDate dob;
    @Column
    private String gender;
    @Override
    public String toString() {
        return "Employee [id= " + id + ", name=" + name + ", department="
                + department + ", dob=" + dob + ", gender="+ gender + "]";
    }
    public Integer getId() {
        return id;
    }
    public void setId(Integer id) {
        this.id = id;
    }
    public String getName() {
        return name;
    }
    public void setName(String name) {
        this.name = name;
    }
    public String getDepartment() {
        return department;
    }
    public void setDepartment(String department) {
        this.department = department;
    }
    public LocalDate getDob() {
        return dob;
    }
    public void setDob(LocalDate dob) {
        this.dob = dob;
    }
    public String getGender() {
        return gender;
    }
    public void setGender(String gender) {
        this.gender = gender;
    }
}
```

15.1.2　创建 DAO 层

创建接口 EmployeeDAO，代码如例 15-2 所示。

【例 15-2】 接口 EmployeeDAO 的代码。

```java
import edu.code.model.Employee;
import java.util.List;
public interface EmployeeDAO {
    List<Employee> get();
    Employee get(int id);
    void save(Employee employee);
    void delete(int id);
}
```

创建类 EmployeeDAOImpl，代码如例 15-3 所示。

【例 15-3】 类 EmployeeDAOImpl 的代码。

```java
package edu.code.dao;
import java.util.List;
import javax.persistence.EntityManager;
import edu.code.model.Employee;
import org.hibernate.query.Query;
import org.hibernate.Session;
import org.springframework.beans.factory.annotation.Autowired;
import org.springframework.stereotype.Repository;
@Repository
public class EmployeeDAOImp implements EmployeeDAO {
    @Autowired
    private EntityManager entityManager;
    @Override
    public List<Employee> get() {
        Session currSession = entityManager.unwrap(Session.class);
        Query<Employee> query = currSession.createQuery("from Employee", Employee.class);
        List<Employee> list = query.getResultList();
        return list;
    }
    @Override
    public Employee get(int id) {
        Session currSession = entityManager.unwrap(Session.class);
        Employee emp = currSession.get(Employee.class, id);
        return emp;
    }
    @Override
    public void save(Employee employee) {
        Session currSession = entityManager.unwrap(Session.class);
        currSession.saveOrUpdate(employee);
    }
    @Override
```

```java
    public void delete(int id) {
        Session currSession = entityManager.unwrap(Session.class);
        Employee emp = currSession.get(Employee.class, id);
        currSession.delete(emp);
    }
}
```

15.1.3 创建 Service 层

创建接口 EmployeeService，代码如例 15-4 所示。

【例 15-4】 接口 EmployeeService 的代码。

```java
import edu.code.model.Employee;
import java.util.List;
public interface EmployeeService {
    List<Employee> get();
    Employee get(int id);
    void save(Employee employee);
    void delete(int id);
}
```

创建类 EmployeeServiceImpl，代码如例 15-5 所示。

【例 15-5】 类 EmployeeServiceImpl 的代码。

```java
package edu.code.service;
import java.util.List;
import edu.code.dao.EmployeeDAO;
import edu.code.model.Employee;
import org.springframework.beans.factory.annotation.Autowired;
import org.springframework.stereotype.Service;
import org.springframework.transaction.annotation.Transactional;
@Service
public class EmployeeServiceImp implements EmployeeService {
    @Autowired
    private EmployeeDAO employeeDao;
    @Transactional
    @Override
    public List<Employee> get() {
        return employeeDao.get();
    }
    @Transactional
    @Override
    public Employee get(int id) {
        return employeeDao.get(id);
    }
    @Transactional
    @Override
    public void save(Employee employee) {
        employeeDao.save(employee);
```

```
    }
    @Transactional
    @Override
    public void delete(int id) {
        employeeDao.delete(id);
    }
}
```

15.1.4 创建 Controller 层

创建类 EmployeeController,代码如例 15-6 所示。

【例 15-6】 类 EmployeeController 的代码。

```
package edu.code.controller;
import java.util.List;
import edu.code.model.Employee;
import edu.code.service.EmployeeService;
import org.springframework.beans.factory.annotation.Autowired;
import org.springframework.web.bind.annotation.DeleteMapping;
import org.springframework.web.bind.annotation.GetMapping;
import org.springframework.web.bind.annotation.PathVariable;
import org.springframework.web.bind.annotation.PostMapping;
import org.springframework.web.bind.annotation.PutMapping;
import org.springframework.web.bind.annotation.RequestBody;
import org.springframework.web.bind.annotation.RequestMapping;
import org.springframework.web.bind.annotation.RestController;
@RestController
@RequestMapping("/api")
public class EmployeeController {
    @Autowired
    private EmployeeService employeeService;
    @GetMapping("/employee")
    public List<Employee> get() {
        return employeeService.get();
    }
    @PostMapping("/employee")
    public Employee save(@RequestBody Employee employee) {
        employeeService.save(employee);
        return employee;
    }
    @GetMapping("/employee/{id}")
    public Employee get(@PathVariable int id) {
        return employeeService.get(id);
    }
    @DeleteMapping("/employee/{id}")
    public String delete(@PathVariable int id) {
        employeeService.delete(id);
        return "Employee removed with id "+id;
```

```
    }
    @PutMapping("/employee")
    public Employee update(@RequestBody Employee employee) {
        employeeService.save(employee);
        return employee;
    }
}
```

15.1.5 修改后端配置文件

修改配置文件 application.properties，代码如例 15-7 所示。

【例 15-7】 修改后的配置文件 application.properties 的代码。

```
spring.datasource.url=jdbc:mysql://localhost:3306/studywebsite?serverTim
ezone=GMT&useUnicode=true&characterEncoding=UTF-8&useSSL=false
spring.datasource.username=root
spring.datasource.password=ws780125
spring.datasource.driver-class-name=com.mysql.cj.jdbc.Driver
spring.jpa.generate-ddl=true
spring.jpa.hibernate.ddl-auto=update
spring.jpa.show-sql=true
spring.jpa.hibernate.ddl-auto=update
server.error.include-message=always
```

15.1.6 数据库文件 studywebsite.sql

创建完数据库 studywebsite 之后，按照数据库文件 studywebsite.sql 创建表 tb_emp 并插入数值，代码如例 15-8 所示。

【例 15-8】 数据库文件 studywebsite.sql 的代码。

```sql
DROP TABLE IF EXISTS 'tb_emp';
CREATE TABLE 'tb_emp' (
  'id' int(11) NOT NULL AUTO_INCREMENT,
  'department' varchar(255) DEFAULT NULL,
  'dob' date DEFAULT NULL,
  'gender' varchar(255) DEFAULT NULL,
  'name' varchar(255) DEFAULT NULL,
  PRIMARY KEY ('id')
) ENGINE=InnoDB AUTO_INCREMENT=8 DEFAULT CHARSET=utf8;
INSERT INTO 'tb_emp' VALUES ('1', 'cs', '1999-10-28', 'male', 'zsf');
INSERT INTO 'tb_emp' VALUES ('2', 'se', '2021-12-03', 'female', 'ls');
INSERT INTO 'tb_emp' VALUES ('3', 'z', '1998-04-02', 'z', 'z');
INSERT INTO 'tb_emp' VALUES ('4', 'jsj', '1998-04-02', 'v', 'zsf');
INSERT INTO 'tb_emp' VALUES ('5', 'ss', '1998-04-02', 's', 'ss');
INSERT INTO 'tb_emp' VALUES ('6', 's', '1998-04-02', 's', 'ss');
INSERT INTO 'tb_emp' VALUES ('7', 'd', '1998-04-02', 'nan', 'zsf');
```

15.1.7 修改后端入口类

修改类 CaseexApplication，代码如例 15-9 所示。

【例 15-9】 修改后的类 CaseexApplication 的代码。

```java
package edu.code;
import org.springframework.boot.SpringApplication;
import org.springframework.boot.autoconfigure.SpringBootApplication;
import org.springframework.context.annotation.Bean;
import org.springframework.web.servlet.config.annotation.CorsRegistry;
import org.springframework.web.servlet.config.annotation.WebMvcConfigurer;
@SpringBootApplication
public class CaseexApplication {
    public static void main(String[] args) {
        SpringApplication.run(CaseexApplication.class, args);
    }
    @Bean
    public WebMvcConfigurer corsConfigurer() {
        return new WebMvcConfigurer() {
            public void addCorsMappings(CorsRegistry registry) {
                registry.addMapping("/**");
            }
        };
    }
}
```

15.1.8 运行后端 Spring Boot 程序

运行入口类 CaseexApplication，成功启动自带的内置 Tomcat。在浏览器中输入 localhost:8080/api/employee 后，效果如图 15-1 所示。

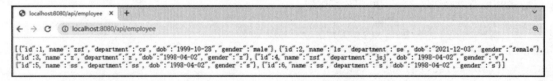

图 15-1　在浏览器中输入 localhost:8080/api/employee 后的运行效果

15.2　React 作为前端的开发

15.2.1　修改文件 App.js 和 App.css

参考 8.2 节用 WebStorm 创建 React 项目 studywebsite，修改项目中 src 目录下文件 App.js，代码如例 15-10 所示。

【例 15-10】 src 目录下文件 App.js 被修改后的代码。

```
import React, { Component } from "react";
```

```
import AddEmployee from "./components/AddEmployee";
import {Route, BrowserRouter as Router, Routes} from "react-router-dom";
import SimpleTable from "./components/Table"; //Table
class App extends Component {
  render() {
    return (
      <Router>
        <Routes>
        <Route exact path="/" element={<AddEmployee/>} />
        <Route exact path="/view" element={<SimpleTable/>} />
        </Routes>
      </Router>
    );
  }
}
export default App;
```

修改项目中 src 目录下文件 App.css，代码如例 15-11 所示。

【例 15-11】 src 目录下文件 App.css 被修改后的代码。

```
.App {
  text-align: center;
}
.App-logo {
  height: 40vmin;
  pointer-events: none;
}
@media (prefers-reduced-motion: no-preference) {
  .App-logo {
    animation: App-logo-spin infinite 20s linear;
  }
}
.App-header {
  background-color: #282c34;
  min-height: 100vh;
  display: flex;
  flex-direction: column;
  align-items: center;
  justify-content: center;
  font-size: calc(10px + 2vmin);
  color: white;
}
.App-link {
  color: #61dafb;
}
@keyframes App-logo-spin {
  from {
    transform: rotate(0deg);
  }
```

```
    to {
      transform: rotate(360deg);
    }
  }
```

15.2.2　创建组件

在项目中 src 目录下创建 components 子目录，在 src\components 目录下创建文件 Table.js，代码如例 15-12 所示。

【例 15-12】 文件 Table.js 的代码。

```
import React from "react";
import { makeStyles } from "@material-ui/core/styles";
                                                    //安装包@material-ui/core
import Table from "@material-ui/core/Table";
import TableBody from "@material-ui/core/TableBody";
import TableCell from "@material-ui/core/TableCell";
import TableContainer from "@material-ui/core/TableContainer";
import TableHead from "@material-ui/core/TableHead";
import TableRow from "@material-ui/core/TableRow";
import Paper from "@material-ui/core/Paper";
import Avatar from "@material-ui/core/Avatar";
import GroupIcon from "@material-ui/icons/Group";//安装包@material-ui/icons
import { Link } from "react-router-dom";
import Typography from "@material-ui/core/Typography";
import CircularProgress from "@material-ui/core/CircularProgress";
const useStyles = makeStyles(theme => ({
  table: {
    minWidth: 600
  },
  avatar: {
    margin: theme.spacing(1),
    backgroundColor: theme.palette.secondary.main
  },
  paper: {
    display: "flex",
    flexDirection: "column",
    justifyContent: "center",
    alignItems: "center",
    margin: `10px`,
    height: "100%",
    width: "99%",
    marginTop: theme.spacing(7)
  },
  link: {
    color: "rgba(0,0,0,0.65)",
    textDecoration: "none",
    marginLeft: "10%",
```

```jsx
      alignSelf: "flex-start",
      "&:hover": {
        color: "rgba(0,0,0,1)"
      }
    }
  }));
export default function SimpleTable() {
  const classes = useStyles();
  const [data, upDateData] = React.useState([]);
  const [firstLoad, setLoad] = React.useState(true);
  let isLoading = true;
  async function sampleFunc() {
    let response = await fetch("http://localhost:8080/api/employee");
    let body = await response.json();
    upDateData(body);
  }
  if (firstLoad) {
    sampleFunc();
    setLoad(false);
  }
  if (data.length > 0) isLoading = false;
  return (
    <div className={classes.paper}>
      <Avatar className={classes.avatar}>
        <GroupIcon />
      </Avatar>
      <Typography component="h1" variant="h5">
        员工信息表
      </Typography>
      {isLoading ? (
        <CircularProgress />
      ) : (
        <TableContainer
          style={{ width: "80%", margin: "0 10px" }}
          component={Paper}
        >
          <Table className={classes.table} aria-label="simple table">
            <TableHead>
              <TableRow>
                <TableCell align="center">姓名</TableCell>
                <TableCell align="center">部门</TableCell>
                <TableCell align="center">性别</TableCell>
                <TableCell align="center">出生年月日</TableCell>
              </TableRow>
            </TableHead>
            <TableBody>
              {data?.map(row => (
                <TableRow key={row.name}>
                  <TableCell align="center">{row.name}</TableCell>
```

```
              <TableCell align="center">{row.department}</TableCell>
              <TableCell align="center">{row.gender}</TableCell>
              <TableCell align="center">{row.dob}</TableCell>
            </TableRow>
          ))}
        </TableBody>
      </Table>
    </TableContainer>
   )}
    <Link className={classes.link} to="/">
      {" "}
      <Typography align="left">
        &#x2190; 返回增加员工信息
      </Typography>{" "}
    </Link>
  </div>
 );
}
```

在项目中 src\components 目录下创建文件 AddEmployee.js,代码如例 15-13 所示。

【例 15-13】 文件 AddEmployee.js 的代码。

```
import React from "react";
import Avatar from "@material-ui/core/Avatar";
import Button from "@material-ui/core/Button";
import CssBaseline from "@material-ui/core/CssBaseline";
import TextField from "@material-ui/core/TextField";
import { Link } from "react-router-dom";
import Grid from "@material-ui/core/Grid";
import GroupIcon from "@material-ui/icons/Group";
import Typography from "@material-ui/core/Typography";
import { makeStyles } from "@material-ui/core/styles";
import Container from "@material-ui/core/Container";
const useStyles = makeStyles(theme => ({
  paper: {
    marginTop: theme.spacing(7),
    display: "flex",
    flexDirection: "column",
    alignItems: "center"
  },
  avatar: {
    margin: theme.spacing(1),
    backgroundColor: theme.palette.secondary.main
  },
  form: {
    width: "100%", // Fix IE 11 issue.
    marginTop: theme.spacing(3)
  },
  submit: {
```

第15章 案例——实现一个简易的员工信息管理系统

```js
      margin: theme.spacing(3, 0, 2)
    },
    textField: {
      marginLeft: theme.spacing(1),
      marginRight: theme.spacing(1),
      width: "100%"
    }
  }));
export default function AddEmployee() {
  const classes = useStyles();
  const [firstLoad, setLoad] = React.useState(true);
  const [selectedDate, setSelectedDate] = React.useState(
    new Date("1998-04-02T21:11:54")
  );
  const [name, setName] = React.useState("");
  const [department, setDepartment] = React.useState("");
  const [gender, setGender] = React.useState("");
  const handleDateChange = date => setSelectedDate(date.target.value);
  const handleNameChange = event => setName(event.target.value);
  const handleDepartmentChange = event => setDepartment(event.target.value);
  const handleGenderChange = event => setGender(event.target.value);
  const [message, setMessage] = React.useState("Nothing saved in the session");
  async function sampleFunc(toInput) {
    const response = await fetch("http://localhost:8080/api/employee", {
      method: "POST",           //*GET, POST, PUT, DELETE, etc.
      mode: "cors",             //no-cors, *cors, same-origin
      cache: "no-cache",        //*default, no-cache, reload, force-cache, only-if-cached
      credentials: "same-origin", // include, *same-origin, omit
      headers: {
        "Content-Type": "application/json"
        // 'Content-Type': 'application/x-www-form-urlencoded',
      },
      redirect: "follow",           //manual, *follow, error
      referrerPolicy: "no-referrer",//no-referrer, *client
      body: JSON.stringify(toInput)    //body data type must match "Content-Type" header
    });
    let body = await response.json();
    console.log(body.id);
    setMessage(body.id ? "Data sucessfully updated" : "Data updation failed");
  }
  const handleSubmit = variables => {
    const toInput = { name, department, gender, dob: selectedDate };
    sampleFunc(toInput);
    setName("");
    setDepartment("");
```

```jsx
      setGender("");
    };
    if (firstLoad) {
      setLoad(false);
    }
    return (
      <Container component="main" maxWidth="xs">
        <CssBaseline />
        <div className={classes.paper}>
          <Avatar className={classes.avatar}>
            <GroupIcon />
          </Avatar>
          <Typography component="h1" variant="h5">
            新增员工信息
          </Typography>
          <form className={classes.form} noValidate>
            <Grid container spacing={2}>
              <Grid item xs={12}>
                <TextField
                  variant="outlined"
                  required
                  fullWidth
                  id="name"
                  value={name}
                  label="姓名"
                  name="name"
                  autoComplete="name"
                  onChange={handleNameChange}
                />
              </Grid>
              <Grid item xs={12} sm={6}>
                <TextField
                  autoComplete="department"
                  name="department"
                  variant="outlined"
                  required
                  fullWidth
                  value={department}
                  id="department"
                  label="部门"
                  onChange={handleDepartmentChange}
                />
              </Grid>
              <Grid item xs={12} sm={6}>
                <TextField
                  variant="outlined"
                  required
                  fullWidth
                  id="gender"
```

```
              value={gender}
              label="性别"
              name="gender"
              autoComplete="gender"
              onChange={handleGenderChange}
            />
          </Grid>
          <Grid item xs={12}>
            <TextField
              id="date"
              label="出生年月日"
              type="date"
              defaultValue="1998-04-02"
              className={classes.textField}
              InputLabelProps={{
                shrink: true
              }}
              onChange={handleDateChange}
            />
          </Grid>
        </Grid>
        <Button
          fullWidth
          variant="contained"
          color="primary"
          preventDefault
          className={classes.submit}
          onClick={handleSubmit}
        >
          保存
        </Button>
        <Grid container justify="center">
          <Grid item>
            <Link to="/view">查看员工信息</Link>
          </Grid>
        </Grid>
      </form>
      <Typography style={{ margin: 7 }} variant="body1">
        状态：{message}
      </Typography>
    </div>
  </Container>
);
}
```

15.2.3 运行前端 React 程序

运行项目 studywebsite，在浏览器中输入 localhost:3000，效果如图 15-2 所示。单击

图 15-2 所示中的超链接"查看员工信息"后,效果如图 15-3 所示。在图 15-2 所示中输入信息后,效果如图 15-4 所示。

图 15-2　在浏览器中输入 localhost:3000 后的运行效果

图 15-3　单击图 15-2 所示中的超链接"查看员工信息"后的效果

图 15-4 在图 15-2 所示中输入信息后的效果

习题 15

实验题

请独立完成本章案例的实现。

附　　录

附录部分以电子版提供，内容涵盖 IDE 安装简介、Spring Boot 应用开发入门、Python 框架应用开发入门、Go 开发入门，请读者扫描下方二维码获取。

电子版附录

参 考 文 献

[1] 王金柱. React.js 16 从入门到实战[M]. 北京: 清华大学出版社, 2020.
[2] 郑均辉, 薛燚. JavaScript+Vue+React 全程实例[M]. 北京: 清华大学出版社, 2019.
[3] 徐超. React 进阶之路[M]. 北京: 清华大学出版社, 2018.
[4] 袁林, 尹皓, 陈宁. React+Node.js 开发实战: 从入门到项目上线[M]. 北京: 机械工业出版社, 2021.
[5] 程墨. 深入浅出 React 和 Redux[M]. 北京: 机械工业出版社, 2017.
[6] 陈屹. 深入 React 技术栈[M]. 北京: 人民邮电出版社, 2016.
[7] 吴胜. Spring Boot 开发实战:微课视频版[M]. 北京: 清华大学出版社, 2019.

图书资源支持

感谢您一直以来对清华版图书的支持和爱护。为了配合本书的使用,本书提供配套的资源,有需求的读者请扫描下方的"书圈"微信公众号二维码,在图书专区下载,也可以拨打电话或发送电子邮件咨询。

如果您在使用本书的过程中遇到了什么问题,或者有相关图书出版计划,也请您发邮件告诉我们,以便我们更好地为您服务。

我们的联系方式:

地　　址:北京市海淀区双清路学研大厦 A 座 714

邮　　编:100084

电　　话:010-83470236　010-83470237

客服邮箱:2301891038@qq.com

QQ:2301891038(请写明您的单位和姓名)

资源下载: 关注公众号"书圈"下载配套资源。

资源下载、样书申请
书圈

图书案例
清华计算机学堂

观看课程直播